POLAR
GEOMORPHOLOGY

INSTITUTE OF BRITISH GEOGRAPHERS
SPECIAL PUBLICATION

No. 4 June 1972

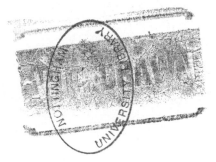

LONDON
INSTITUTE OF BRITISH GEOGRAPHERS
I KENSINGTON GORE SW7

1972

Printed in Great Britain by Alden & Mowbray Ltd
at the Alden Press, Oxford

Contents

Preface   *page* v

The role of thermal régime in glacial sedimentation *by* G. S. BOULTON   I

The wear of sandstone by cold, sliding ice *by* R. HOPE, H. LISTER and R. WHITEHOUSE   21

Deglaciation of the Scoresby Sund fjord region, north-east Greenland *by* S. FUNDER   33

The contribution of radio echo sounding to the investigation of Cenozoic tectonics
and glaciation in Antarctica *by* D. J. DREWRY   43

Volcanic record of Antarctic glacial history: implications with regard to Cenozoic
sea levels *by* W. E. LE MASURIER   59

Evidence from the South Shetland Islands towards a glacial history of West Antarctica *by* B. S. JOHN   75

Tors, rock weathering and climate in southern Victoria Land, Antarctica *by* E.
DERBYSHIRE   93

Valley asymmetry and slope forms of a permafrost area in the Northwest Territories, Canada *by* B. A. KENNEDY and M. A. MELTON   107

Modification of levee morphology by erosion in the Mackenzie River delta, North-
west Territories, Canada *by* D. GILL   123

Relationships between process and geometrical form on High Arctic debris slopes,
south-west Devon Island, Canada *by* P. J. HOWARTH and J. G. BONES   139

Processes of soil movement in turf-banked solifluction lobes, Okstindan, northern
Norway *by* C. HARRIS   155

The nature of the ice-foot on the beaches of Radstock Bay, south-west Devon
Island, N.W.T., Canada *by* S. B. McCANN and R. J. CARLISLE   175

The solution of limestone in an Arctic environment *by* D. I. SMITH   187

Processes of solution in an Arctic limestone terrain *by* J. G. COGLEY   201

The British Geomorphological Research Group   213

The Institute of British Geographers   214

Publications in print   215

Preface

The fourteen papers in this volume have been contributed by members of the British Geomorphological Research Group, following a Symposium on Polar Geomorphology held at the University of Aberdeen, 4 and 5 January 1972, during the Annual Conference of the Institute of British Geographers. Dr R. J. Price and Dr D. E. Sugden shared the work of compiling the volume and organizing the Symposium, and undertook the correction of proofs; to them the Editors express their thanks. Some of the discussion that took place at the Symposium has been included where this seemed worthy of record.

C. EMBLETON, Hon. Editor
C. E. EVERARD, Hon. Assistant Editor

The role of thermal régime in glacial sedimentation

G. S. BOULTON

School of Environmental Sciences, University of East Anglia

Revised MS received 24 November 1971

ABSTRACT. The thermal régime of a glacier, determined by climate, mass balance and ice thickness, can be used to define four boundary conditions at a glacier sole. These are: a zone (A) of net basal melting, a zone (B) in which there is a balance between melting and freezing, a zone (C) in which sufficient meltwater freezes to the glacier sole to maintain it at the melting point, and a zone (D) in which the glacier sole is below the melting point. In zone A, the glacier slips over its bed and erosion takes place by abrasion and crushing. The products of erosion are transported in a thin basal regelation layer no more than 0·5–1·0 m thick, and are subsequently deposited as lodgement till in zones where the frictional retardation against the bed is high. Subglacial materials are readily deformed because of high subglacial water pressures. In zone B, processes are essentially similar, but rates of erosion and deposition may well differ because of lower subglacial water pressures. In zone C, the glacier still slips over its bed, but plucking of sub-glacial material occurs in addition to abrasion and crushing. The products of erosion may be carried up to high levels in the ice by net freezing of water at the glacier sole. Because of the high level of transport, subsequent deposition tends to take the form of flow till and melt-out till rather than lodgement till. In zone D, the glacier adheres to its bed, there is no basal slip and movement takes place by internal flow alone. Abrasion and crushing are relatively unimportant processes, but plucking of very large erratics may well occur. Again there is little deposition of subglacial lodgement till, debris carried above the glacier sole being mainly deposited as melt-out till and flow till. Deformation of frozen subglacial sediments does not occur. In all these cases, the major variables affecting process are the roughness of the bed, the discharge of subglacial water, and the permeability of subglacial strata.

THERE is a growing body of observations on contemporary processes of glacial sedimenta-tion. In this paper, it is suggested that the thermal régime of a glacier at any one point determines the boundary conditions at the glacier sole. These define the sedimentational processes which can occur there and thus can be used to explain spatial variations in these processes. Given these thermal boundary conditions, specific processes of erosion, trans-port and deposition depend upon the configuration of the glacier bed, the materials which compose it, ice velocity and ice thickness. These latter assertions will be justified in detail in a subsequent paper. It is possible to predict the variations in sedimentational processes beneath known glaciers, and also to construct convenient models of ancient glaciers and ice sheets for which such variations can be observed.

Thermal régime itself is largely determined by mass balance and surface temperature, the nature and speed of glacier movement, and ice thickness.

THERMAL BOUNDARY CONDITIONS

The heat balance at a glacier sole is relatively simple. Sources of heat are the geothermal heat flux, the heat produced by glacier sliding, and heat derived from the glacier surface. Let H_{geo} = geothermal heat flux

S = speed of glacier sliding

τ = average shear stress at the glacier bed

I

Ts = temperature at the glacier surface below the depth of seasonal temperature fluctuation

Tb = temperature at the glacier bed

C = thermal conductivity of ice

h = glacier thickness

k = conversion factor, energy to heat units

b = specific heat of ice

a = net budget

y = arbitrary height above the glacier bed

H_{geo} can be regarded as constant at any one locality, and has an average value of 167 J cm^{-2} yr^{-1}; the heat produced by glacier sliding is equal to $\dfrac{S\tau}{k}$; and the temperature gradient within the glacier just above the glacier sole is equal to $\dfrac{(Tb-Ts)}{h^*}$ (G. de Q. Robin, 1955); the sign may be positive, indicating an upward gradient, or negative indicating a downward gradient. h^* is a 'compensated' ice thickness which serves to linearize the gradient.

$$h^* = \int_0^h \exp\left(\frac{-bay^2}{2Ch}\right) dy$$

The temperature gradient just above the glacier sole determines how much heat can enter the glacier per unit time.

Assuming a horizontal bed, three boundary conditions can initially be defined:

$$\frac{S\tau}{k} + H_{geo} > \frac{(Tb-Ts)}{h^*}C \tag{1}$$

$$\frac{S\tau}{k} + H_{geo} \approx \frac{(Tb-Ts)}{h^*}C \tag{2}$$

$$\frac{S\tau}{k} + H_{geo} < \frac{(Tb-Ts)}{h^*}C \tag{3}$$
$$\text{(J. Weertman, 1961)}$$

In the first case more heat is provided at the glacier sole than can be conducted upwards through the glacier, and therefore net melting of basal ice occurs. The glacier sole is at the melting point and it slides over its bed.

In the second case, the heat provided at the glacier sole per unit time is approximately equal to the amount which can be conducted through the glacier per unit time. Thus there is no excess basal melting, but no excess freezing, and the glacier slides over its bed.

In the third case, the amount of heat provided at the glacier sole is insufficient to prevent freezing. Thus water in the pores of subglacial rocks may be frozen to some depth beneath the sole. One of the effects of a glacier being frozen to its bed is to prevent sliding along this interface. There is evidence of strong interfacial adhesion between ice and rock at temperatures below the melting point, so that shear takes place within the ice itself (D. Tabor and J. C. F. Walker, 1970; H. H. G. Jellinek, 1959). Forward movement of the glacier is thus by internal flow alone. Subglacial water is not produced by the glacier, any free water near the glacier sole being liable to freeze.

These three boundary conditions can be thought of as defining three thermal zones, each of which has its own attributes as far as subglacial processes are concerned. The three most important variables in defining these zones are sliding velocity (S), ice thickness (h), surface temperature (Ts) and net budget (a). The larger these variables, the more likely it is that basal melting will occur, basal freezing being favoured by small values. In any large glacier these variables may change considerably from one area to another, and thus basal conditions themselves may vary considerably beneath the different parts of a single glacier.

VARIATIONS IN THERMAL RÉGIME OF CONTEMPORARY GLACIERS

On the Antarctic ice sheet and in the middle of the Greenland ice sheet, summer temperatures do not rise above 0°C. Thus there is no meltwater and accumulated snow remains dry at temperatures well below zero. Seasonal fluctuations in temperature are attenuated at shallow depths as dry snow has a relatively poor thermal conductivity.

In the accumulation areas of many circumpolar glaciers such as those of Spitsbergen (H. V. Sverdrup, 1935) and Sukkertoppen Ice Cap in West Greenland (F. Loewe, 1966), temperatures may rise above 0°C for one or two months of the summer. The resultant meltwater percolates down into the underlying ice until it freezes. Thus heat is transported down into the newly accumulated snow by advection which is relatively efficient compared with solid conduction. Thus the summer, 'warm', component of seasonal temperature change has a much greater effect on ice temperatures below the limit of seasonal fluctuations in temperature, producing temperatures very close to the melting point. In the accumulation zones of glaciers in such areas as the Alps and Norway, the summer warm period is longer and advection of heat by percolating meltwater is even more important. In these glaciers the temperature of ice in the accumulation area is rarely below the pressure melting point.

In the ablation zones, transfer of heat through the ice is by conduction alone. Thus in polar glaciers such as those of Baffin Island, Spitsbergen and Greenland, summer temperatures are not overemphasized by an advection component of heat transfer, and ice below the depth of seasonal fluctuation tends to be below the melting point. In glaciers in temperate areas, summer temperatures and average annual temperatures in the ablation zones tend to be high, and therefore conduction alone is capable of maintaining ice at the melting point.

Given this kind of variation in surface temperature, let us superimpose the effect of ice thickness. Although surface temperatures are very low in the central part of the Antarctic ice sheet (less than -40°C), the thickness of ice is very great, and thus basal melting is a distinct possibility (*cf.* Weertman, 1966). Basal melting is also likely beneath the accumulation zones of many circumpolar ice caps where summer melting is important. In the ablation zones of these same ice caps the ice tends to be cold and basal freezing is likely (V. Schytt, 1969); the subglacial layer of frozen sediments is essentially a subglacial extension of the permafrost. Basal melting also occurs beneath temperate glaciers. There the ice is at the pressure melting point throughout and, as pressures increase with depth, so must temperatures decrease. Such a negative, downward temperature gradient would automatically produce basal melting.

Figure 1 is a schematic diagram showing the ways in which these thermal zones can be combined to illustrate the variety of thermal conditions that might exist within one glacier. One important addition needs to be made. Water, actively being produced at a glacier sole,

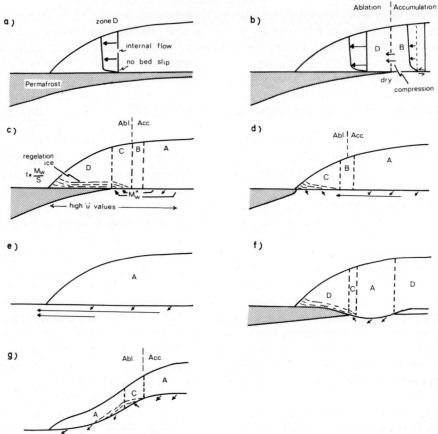

FIGURE 1. Ways in which different thermal zones can be combined. Each zone has its own characteristic processes of erosion, transport, deposition and sediment deformation. The patterns in which zones are combined also influence processes at any one point

cannot accumulate in that position. Clearly it must flow away down a pressure gradient which is largely that of decreasing ice thickness. Away from subglacial channels, it may flow through underlying beds if these are sufficiently permeable or sufficiently thick. But if the hydraulic conductivity of these beds is inadequate it must flow as a thin water layer between glacier sole and bed.

Weertman (1966) has shown the hydraulic gradient beneath a glacier to be

$$\frac{\rho_i g \dfrac{dh}{dx}}{\mu} \tag{4}$$

where ρ_i = density of ice
g = acceleration owing to gravity
$\dfrac{dh}{dx}$ = slope of glacier surface
μ = viscosity of water

The hydraulic conductivity (U^*) of subglacial beds is

$$U^* = KD^* \frac{\rho_i g \frac{dh}{dx}}{\mu} \qquad (5)$$

where K = permeability
D^* = thickness of permeable beds

If the total volume of meltwater produced up-glacier of the beds in question per unit time is Mw^*, then unless $U^* \geqslant Mw^*$, there must be a discharge of

$$Q = Mw^* - KD^* \frac{\rho_i g \frac{dh}{dx}}{\mu} \qquad (6)$$

in a water layer between glacier sole and bed. The discharge in the water layer is related to its thickness

$$Q = D^3 \frac{\rho_i g \frac{dh}{dx}}{12\mu} \qquad (7) \text{ (Weertman, 1966)}$$

where D = thickness of the water layer.
Combining (6) and (7), the thickness of the water layer is

$$D = \frac{12\mu Mw^*}{g \cdot \frac{dh}{dx}} - 12KD^{*\frac{1}{3}} \qquad (8)$$

The existence of such a water layer is of fundamental importance in considering processes of erosion, transport and deposition, in that the effective stress between particles embedded in the glacier sole and the bed is not simply ice overburden but the difference between ice overburden and water pressure ($\rho_i gh - u$, where u = water pressure). It is possible that, under certain conditions, the effective pressure will be very small, even beneath very large ice overburdens. For instance, the water pressure beneath an overburden of 250 m of ice (25 bars) of the Glacier d'Argentière was 14 bars (R. Vivian, 1970), giving an effective pressure of only 11 bars. Because the existence and size of the water layer depends on the hydraulic conductivity of underlying beds, the effective pressure will be smaller where the glacier passes over impermeable beds and larger where it passes over permeable beds (Fig. 2).

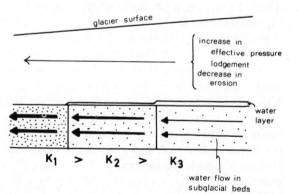

FIGURE 2. The effect of the permeability (K) of subglacial beds on the thickness of a basal water layer. The thickness of the water layer controls effective pressure between ice and bed, and thus affects erosion and deposition

Consider a glacier such as that shown in Figure 1c. Water would move from an internal zone of basal melting into an outer zone of basal freezing. The downward extension of permafrost beneath the glacier would effectively prohibit its further outward movement although high pore-water pressures would develop beneath the permafrost. If a steady state is to be maintained and the centre is not to become unstable by continued accumulation of subglacial meltwater, this meltwater must be removed from the system by freezing to the glacier sole on the up-glacier side of the outer cold zone. Freezing of this water to the glacier sole would raise the basal temperature, thus producing an intermediate zone in which the glacier sole is maintained at the melting point, and thus able to slide, as long as sufficient meltwater is continuously available. Basal slip and the net acquisition of ice by the glacier sole will continue as long as

$$ Mw = \frac{\frac{(Tb-Ts)}{h}C - H_{geo} + \frac{S}{K}}{L} \tag{9} $$

where Mw = mass of water freezing to the glacier sole per unit area, and L = latent heat of fusion of ice. The total amount of water freezing to the glacier sole in the zone must equal the amount produced in the inner zone of melting (Mw^*). The thickness of ice produced in this zone would be $\frac{Mw^*}{S}$.

It is now possible to differentiate several types of glacier on the basis of thermal régime (Fig. 1). First, some glaciers are composed of cold ice throughout, in which the sole is below the melting point, and in which no basal slip occurs. Effectively, permafrost extends beneath the whole of the glacier (Fig. 1a). Secondly, there are dominantly cold glaciers in which the melting point isotherm reaches the bed in the central parts of the glacier. Basal slip occurs there but there is no excess melting (Fig. 1b). A third type of glacier exhibits excess basal melting in the middle of the accumulation zone. Water moving outwards from this zone under a hydrostatic pressure gradient then freezes to the bed beneath cold ice in an intermediate zone; thus, new basal ice is acquired and bed slip occurs. In a further outer zone, insufficient water is provided at the glacier sole to retain it at the melting point; it thus freezes to its bed, no basal slip occurs and the melting point isotherm lies at some depth beneath the glacier sole. Many circumpolar glaciers probably approximate to this model (Fig. 1c). Fourthly, there are glaciers similar to those above but in which the first two zones only occur, there being no substantial extension of permafrost beneath the glaciers, although it may occur widely beyond their margins (Fig. 1d). In this case, not all of the water provided in the internal zone need be removed by freezing in the outer zone. Fifthly, there are glaciers in which basal melting alone occurs. Basal slip occurs throughout (Fig. 1e).

The above cases refer only to glaciers flowing over flat beds. Irregular beds or strikingly different climatic régimes in different parts of the glacier may produce much more complex patterns of thermal variation. Thus, a deep trench beneath what would otherwise be a cold part of a large ice cap or ice sheet may induce melting at the glacier sole (Fig. 1f) and the erosional and sedimentological processes associated with such conditions. Similarly, a glacier flowing rapidly from a cold high mountain area into a temperate zone may be composed of an up-glacier cold zone and a down-glacier temperate zone (Fig. 1g).

PROCESSES OF GLACIAL SEDIMENTATION

It is proposed that the nature of the processes of subglacial erosion, glacial transport, deposition, and post-depositional deformation which can occur at any one place are determined by the four thermal boundary conditions defined above. These processes are considered below in order.

Subglacial erosion

Zone A: net basal melting (equation 1) Subglacial erosion can be effected both by subglacial streams and by ice and the particles embedded in it. The latter, glacial erosion *sensu stricto*, is generally thought to comprise crushing by particles embedded in the ice, polishing by these same particles and perhaps even the ice surface itself, and plucking by which loose subglacial blocks around which ice has formed adhere to the glacier sole. Crushing and polishing can be readily seen in tunnels beneath temperate glaciers in which there is net melting of the glacier sole, and are most effectively accomplished by larger particles. They produce striations and smoothing of bedrock where the glacier sole conforms well to the shape of the bed. Crushing of bedrock is more effective under lesser thicknesses of ice where subglacial cavities are numerous, for in such circumstances bedrock subjected to crushing stresses can expand laterally and produce considerable amounts of debris. In confined situations, the effective crushing strength of bedrock is very much increased. For instance, it is suggested that the steep lee-side faces of *roches moutonnées*, formed beneath ice which is melting at its base, are not the product of plucking by the freezing of meltwater in joints as has been commonly supposed (e.g. H. Carol, 1947) but form because of the possibility of lateral expansion of the rock into lee-side cavities. They are thus likely to form beneath small thicknesses of ice. It is also suggested that plucking, which requires freezing beneath the glacier sole, is relatively unimportant in zones of net basal melting.

The crushing and polishing components of erosion depend intimately on the value and distribution of shear stresses at the glacier bed and the normal load across particles in contact with the bed. In a glacier with considerable basal melting, relatively impermeable subglacial beds will induce the formation of a water layer between sole and bed ($Mw > U^*$), effectively drowning the smaller obstacles to flow and thus reducing the basal shear stress over wide areas, although increasing it over larger obstructions. Similarly the normal load ($\rho_i gh$) will be reduced to an effective pressure ($\rho_i gh - u$). If the effective pressure were to increase, the pressure at the contact between particles embedded in the sole and the glacier bed would increase, and thus more crushing and abrasion would be expected. However, consider a particle sliding over a bedrock surface. The particle will move over the bed if the shear stress (τ) exerted by the ice against the particle is

$$\tau \geqslant (\rho_i gh - u)F$$

where F is a coefficient of friction. If, however,

$$(\rho_i gh - u) \geqslant \frac{\tau}{F} \tag{10}$$

particles will not move over the bed. They thus cease to erode the bed and the ice flows around them. Thus, in order that abrasion and crushing should occur under greater ice thicknesses a high pressure in a basal water layer is required (Fig. 2). Abrasive and crushing power increase to a limiting value of ice thickness, beyond which high water pressures in a basal water layer are required to maintain this erosion. If thick subglacial beds are highly permeable, a water layer will not form, and crushing and abrasion will be reduced.

The fact that large quantities of water can be produced beneath certain glaciers, and that this water, together with surface-derived water, can survive in the liquid state, makes considerable subglacial fluvial erosion a possibility. If subglacial water finds its way into large channels with large hydraulic gradients, rapid rates of flow will result. Vivian (1970) has drawn attention to the possibility that subglacial stream erosion might be more important than glacial erosion *sensu stricto* in glaciers with excess basal melting. He compares fluvioglacial rates of erosion amounting to 2 cm yr^{-1} and high sediment discharges by subglacial streams, with the relatively small discharges of basally derived debris transported by these glaciers (p. 248). This suggests the possibility that a large part of the excavation of U-shaped valleys by certain temperate glaciers may result from subglacial fluvial erosion.

Zone B: balance between melting and freezing (equation 2) The basic processes are likely to be similar to those in zone A. If the whole glacier lies within zone B, there will be less water at the bed and effective stresses will be greater, reducing crushing and abrasion under greater ice thicknesses. If this zone, however, lies down-glacier of a zone A, high water pressures could again occur.

Zone C: water freezes to moving glacier sole (equation 9) In this case, the glacier sole slips over its bed because the freezing to it of meltwater that has flowed from adjacent zones of net basal melting maintains the glacier sole at the melting point. Thus, abrasion and crushing are likely to be as important as in the case of ice with net basal melting but, in addition, the process of plucking is likely to occur. Loose or heavily jointed materials lying at the glacier bed will tend to adhere to the glacier sole as ice forms around them or in contact with them.

Zone D: subglacial materials frozen (equation 3) In this case the adhesive strength of the contact between the glacier sole and the frozen subglacial bed is greater than the shear stress across this contact, and therefore no slip occurs between the two. Differential movement does, however, occur within the ice above this interface and material can therefore be transported entirely englacially. It seems most plausible that the level above which such movement will take place will be defined by a set of smooth flow lines approximating to the shape of the bed. Thus, in Figure 3, movement does not occur at the glacier/bed interface but along a smooth curve which lies just above the prominences on this interface. The lowest level at which debris can be transported will coincide with such a flow line. Moving debris is only likely to come into contact with the glacier bed, and therefore erode it, at the summits of protuberances although, even here, slip need not occur between the glacier sole and its bed. The relative ineffectiveness of the processes of crushing and abrasion where the

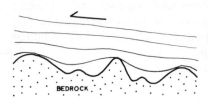

FIGURE 3. Plastic flow lines above an irregular glacier bed in which there is no slip at the ice/bed contact

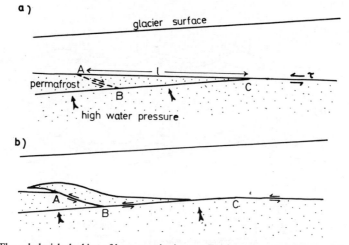

FIGURE 4. The subglacial plucking of large erratics in zone D where the subglacial permafrost is thin

sole of a moving glacier adheres to a frozen bed is illustrated at the margins of the Greenland ice sheet near Thule where boulders overridden by ice have not even been deprived of
their coating of lichen (R. P. Goldthwait, 1960).

However, where the melting-point isotherm does not lie far beneath the glacier bed,
there exists the possibility of quarrying or plucking of very large erratics by the glacier. If
the depth of freezing in subglacial materials coincides with major joints, or planes of weakness along which the frictional resistance is reduced by the high pore-water pressures likely
to occur at this point, then the material may be carried forward by the glacier as long as
there is no obvious hindrance to forward movement of the block (Fig. 4). The newly
quarried mass would then move up into the ice along a flow line such as one of those
indicated in Figure 3.

In Figure 4, assuming the frozen mass to be rigid, the disrupting force is the shear
stress at the glacier bed. For shear to take place along the line ABC, this would be $\tau.l$,
where l is the distance AC. The strength of materials along the shear plane is made up of
two components: (1) the strength of frozen sediments between A and B, F.AB, where F is
the shear strength of the frozen sediments, and (2) the frictional resistance along BC,
$(N-u) \tan \phi$.BC, where N is the average normal load across BC, assuming no cohesion.
Fracture will occur if

$$\tau l \geqslant F.\text{AB} + (N-u) \tan \phi \ \text{BC}$$

As values of F are likely to be larger than τ, fracture is more likely if l/AB is large. High
values of pore-water pressure will also assist fracture by reducing the first strength term.

Such a mechanism could effect the plucking of many known large erratics of unlithified or lithified materials, for instance, the massive erratics (up to 0·5 km in length) on the
Norfolk coast composed of lower Pleistocene sands, silts and clays, and large masses of
chalk. Moreover, this form of plucking need not only operate at the margin of zone D. It
could occur where the depth of subglacial freezing is relatively shallow. In addition,
erratics incorporated by this process might well be subsequently broken up during transport. Thus, under suitable subglacial conditions, relatively large volumes of debris could
be eroded and incorporated in zone D.

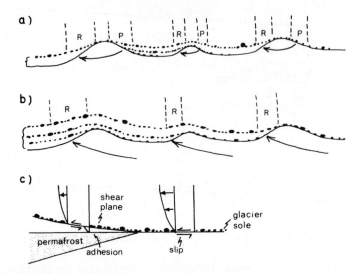

FIGURE 5a. The thin debris–charged regelation layer at the base of temperate ice. Pressure melting (P) takes place on the up-glacier flanks of obstructions, and regelation (R) of the resultant meltwater (marked by an arrow) takes place on the down-glacier flank.
b. The net acquisition of basal debris-charged ice by glaciers in zone C. Thick debris sequences can thus accumulate.
c. The change from zone C (or B) to zone D, involving cessation of bed slip. The resultant compression gives rise to thrust planes along which basal ice may be transported to a higher level

Various opinions have been expressed about the effectiveness of glacier erosion in different areas. D. L. Linton (1959) suggested that ice had been unable to erode effectively on the high plateaux of the Cairngorms; J. B. Bird (1959) argued similarly for large areas of Canada, and V. Tanner (1938) for parts of Finland. There are other areas in which glacial erosion has been very effective, especially lowland areas (C. Embleton and C. A. M. King, 1968). These varying opinions do not necessarily conflict. On the basis of the theory presented here, the intensity of glacier erosion would vary with thermal régime. For instance, under certain circumstances, a greater erosional effect with increased ice thickness (tendency towards melting) would be expected, which is quite compatible with the varied opinions referred to above.

Transport of subglacially-derived debris

Zone A: net basal melting (equation 1) The transport of subglacially derived debris at the base of temperate ice has been described from several areas (e.g., B. Kamb and E. LaChapelle, 1964; J. G. McCall, 1960). One of the ways in which the products of erosion may find their way into the basal ice was described by Kamb and LaChapelle (1964). Water derived from pressure melting on the up-glacier flanks of obstructions refroze on the down-glacier flank beneath debris embedded in the glacier sole (Fig. 5a). Such mass transfer is common beneath temperate glaciers and the basal regelation ice so formed is often heavily charged with debris. Debris-charged regelation ice formed behind one obstruction tends to be destroyed by pressure melting in front of subsequent similarly-sized obstructions. The thickness of the regelation layer is limited because the rate of pressure melting, and the similar rate of formation of regelation ice, is controlled by the rate at which heat flows through the obstacle. Sufficiently large heat fluxes are only possible where obstacles are small (of the order of several centimetres), and therefore the thickness of the regelation layer is similarly small.

Another mechanism by which debris may be incorporated is the interaction between particles embedded in the ice and between these particles and the bed, interactions which are responsible for limited amounts of dispersion in the basal ice. The limited vertical mobility conferred by these mechanisms tends to restrict debris to a thin basal layer in the ice, rarely extending for more than 0·5—1 m above the glacier sole (Fig. 5a). The super-imposition on this system of general basal melting means that debris introduced into an englacial position over a series of obstacles will gradually descend to lower levels if the bed is smooth in the lee of the obstacles.

Where cavities occur, general basal meltwater produced at a melting point below 0°C will tend to freeze when it flows into a cavity, thus producing greater thicknesses of regelation ice (as much as 1 m has been observed by the author).

Zone B: net balance between melting and freezing (equation 2) In this case the mode of incorporation will be basically similar to the previous case, except that the absence of net basal melting will tend to produce a basal debris layer of constant thickness if the ice flows over a smooth bed.

Zone C: water freezes to moving glacier sole (equation 9) In this zone, subglacial water must freeze to the moving glacier sole to maintain it at the melting point. If this water moves as a thin layer between glacier sole and bed, Weertman (in press) has shown that flow will tend to be concentrated at the low points of the bed, and therefore most of the freezing of ice will take place at such points. If water is provided largely from highly permeable subglacial beds, it will flow most readily to those depressions in the glacier bed where the ice exerts least pressure, where it will freeze to the sole. Thus, by both processes, ice will tend to freeze to the glacier sole at low points of the bed, so forming continuous bands of ice beneath the debris-charged glacier sole as it moves over the crests of obstructions (Fig. 5b and G. S. Boulton, 1970). The thickness of the debris-rich ice so formed will be $\frac{Mw^*}{S}$.

Zone D: subglacial materials frozen (equation 3) Incorporation in this zone will be restricted to frozen material quarried in the way described on p. 9, although material acquired in other up-glacier zones will of course be carried in the ice flowing above the stationary sole.

The junction between an inner zone in which the ice is slipping over its bed, and an outer zone that is frozen to its bed, is potentially interesting. Commonly, as much as half the forward motion of a glacier can be accounted for by bed slip and half by internal flow. The transition to a zone in which bed slip does not occur will lead to considerable longitudinal compression and relatively high stresses which will tend to generate discrete thrust planes. Thrusting does not appear actually to incorporate subglacial material, but merely moves already englacial debris to a higher level (Boulton, 1970). Thus, if a glacier sole frozen to its bed were succeeded up-glacier by a zone in which there was a balance between melting and freezing, the basal regelation layer, rich in debris, would be forced up into the ice above a thrust plane (Fig. 5c). If, however, the contact between the two zones were moving in an up-glacier direction during glacier retreat or climatic deterioration, a series of sequential thrust planes would form, producing a series of englacial debris bands.

Even in those glaciers in which basal melting normally occurs, winter cooling produces a cold ice wedge at the extreme margin. The annual formation of such a cold ice wedge could well produce sporadic incorporation by freezing in the manner described by Weertman (1961). The depth at which this could take place would be limited by the attenuation of the winter cold wave and would be of the order of 10 m. J. L. Anderson and J. L. Sollid (1971) appear to have observed this process at the margin of Midtdalsbreen in Norway.

Deposition of till

Restricting discussion initially to ice caps or ice sheets in which subglacially-derived debris alone occurs, till may be deposited subglacially from active ice, supraglacially as ablation till, or by the melting-out of stagnant masses of buried ice. The relative importance of these three modes of deposition clearly depends upon the position in which debris is transported. If debris is transported immediately above the glacier sole, subglacial lodgement is most likely to occur. If debris is transported some distance above the sole, considerable basal melting must occur before lodgement is effected; otherwise large obstructions must project into the ice up to the level of transport in order to arrest movement. The greater part of till is therefore likely to be deposited as ablation till and flow till.

Thus the mode of deposition of debris as till at any point depends on the position in which debris is transported, which itself depends on thermal régime. The details of lodgement depend on additional factors, some of which are again dependent on thermal régime. The deposition of flow till and melt-out till have been described previously (Boulton, 1970, 1971).

Zone A: net basal melting (equation 1) Because of the predominance of basal melting and the concentration of debris transport immediately above the bed, deposition in this zone will largely be of subglacial lodgement till. It is considered that subglacial lodgement occurs when the frictional drag on particles of debris being moved over the bed equals the tractional force exerted on it by the glacier ice, an assertion that will be justified in a subsequent publication. Frequent contacts between particles of debris and the bed will tend to produce a high rate of lodgement in this zone. From equation (10), the frictional drag will increase if the coefficient of friction increases (rough bed) or if the overburden increases, and thus one would expect the rate of deposition to be greater under greater ice thicknesses. However, the frictional drag may be reduced if there is a water layer between glacier sole and bed, for this would lubricate the sole/bed interface and reduce the effective normal pressure to $\rho_i gh - u$. The absence of a water layer owing to highly permeable subglacial beds will therefore tend to increase the rate of deposition (Fig. 2). Thus lodgement will be favoured under thicker ice by areas where the bed is rough or where it is composed of highly permeable sediments. Deposition will initially be concentrated against obstructions to form drumlinoid features, but will eventually tend to fill up low points which, together with erosion of protuberances, will tend to produce a less irregular bed.

If this zone (A) lies down-glacier of a zone in which a thick englacial debris sequence is being included (e.g. zone C), basal melting could lower this debris into contact with the bed against which lodgement might take place.

The distribution of areas of lodgement in zone A will depend on the presence of debris within the ice, the frictional characteristics of the transported debris, the roughness of the bed, the permeability of the bed, subglacial water pressures and ice velocity.

Zone B: net balance between melting and freezing (equation 2) If no meltwater is introduced into this zone, the rate of subglacial lodgement is likely to be higher than in zone A, but if meltwater is present a similar pattern of lodgement is likely to occur.

Zone C: freezing of water to moving glacier sole (equation 9) Although there is no reason why lodgement should not take place in this zone, net freezing at the glacier sole is likely to re-incorporate many of the products of lodgement, and therefore net erosion by plucking of the bed will occur. Any thick englacial debris sequence built up in this zone is unlikely to be deposited at the base of the glacier because there is no melting at that interface. Englacial debris at a relatively high level of transport will encourage dominant supraglacial deposition if basal freezing extends to the glacier margin. Surface melting will release englacial debris supraglacially as ablation till (Fig. 1d) where, if sufficiently fluid, it may give rise to flow till and the series of characteristic landforms associated with flow tills. This generally occurs if the glacier is retreating, but if the glacier is advancing, it normally has a much steeper front, debris falls from the terminal cliff and is overridden by the advancing glacier (E. J. Garwood and J. W. Gregory, 1898). Thus subglacial till may form by recycling *via* the upper surface of the glacier.

Zone D: subglacial materials frozen (equation 3) In the absence of water at the glacier sole and of basal slip, deposition of lodgement till at the base of the glacier will be much less important than the deposition of ablation till and flow till at its upper surface (Fig. 1c). If lodgement does occur, it is most likely to be against larger obstructions which project into the ice and intercept debris-carrying flow lines. Thus drumlins might be a predominant feature of lodgement beneath ice which is frozen to its bed. One effect of lodgement around such obstructions is that the basal ice on the lee side will have been depleted of debris which it is unlikely to regain unless quarrying, such as is described on p. 9, occurs.

DEFORMATION OF SUBGLACIAL MATERIALS

These materials may be pre-existing sediments such as proglacial outwash, or earlier tills over which the glacier has advanced, or they may be lodgement tills recently deposited by the glacier. The deformation may be of two types, plane strain in a confined situation under a shear stress imposed by the overriding glacier, or flow in response to local differential stresses, for instance, flow of wet till into a cavity beneath the glacier.

Deformation in response to the shear stress imposed by the overriding glacier

The average shear stress at the bed of a glacier can be approximated by $\tau = \rho_i g h \sin \alpha$ where α = surface slope of glacier. This is probably true in detail for areas of the bed composed of small obstacles with regular spacing. The average value of τ can be regarded as constant for all except very small values of h and is of the order of $1 - 1 \cdot 5$ bars. Apart from bedrock, two types of unlithified sediments underlie the glacier sole. First, there are those sediments which are frozen (zone D) and which have a relatively high shear strength, much higher than the stress imposed by the glacier, and are therefore unlikely to be deformed by the glacier. Secondly, there are unfrozen sediments lying beneath the glacier sole, in the pores of which there may be appreciable quantities of water (zones A, B, C).

The shear strength of unlithified materials immediately below the glacier sole can be approximated by the Coulomb-Terzaghi equation:

$$\sigma = C + E \tan \phi \quad \text{for cohesive materials (till, clay etc.)}$$
$$\sigma = E \tan \phi \quad \text{for non-cohesive materials}$$

where σ = shear strength
  C = cohesion
  E = effective normal stress

Therefore, materials beneath the glacier sole will deform when

$$C + E \tan \phi \leqslant 1\text{--}1 \cdot 5 \text{ bars}$$
$$\text{or } E \tan \phi \quad \leqslant 1\text{--}1 \cdot 5 \text{ bars}$$

The strength of subglacial materials beneath a certain load of ice will be determined by the pore-water pressure in the sediment, which controls effective pressure ($E = \rho_i g h - u$). High pore-water pressures are likely to occur beneath zones A, B and C, and observations beneath actual glaciers show these to exist. Pressure values of 14 bars have been measured by Vivian (1970) beneath the Glacier d'Argentière in the French Alps, and peak values of about 13 bars (steady values of about 5 bars) have been measured by W. H. Mathews (1964) beneath South Leduc Glacier in British Columbia.

Figure 6a shows typical values for the shear strengths of cohesive and non-cohesive materials for different values of ice thickness and zero pore-water pressure. Deformation of gravel such as A will not occur with ice thicknesses greater than 25 m, of such sand as B for thicknesses greater than 35 m, and of clay such as D for thicknesses greater than 70 m. Thus, in dry conditions, it is only beneath the glacier margin that deformation will take place. Figure 6b shows the pore-water pressures required in different subglacial sediments beneath different thicknesses of ice in order to allow sediments to be deformed immediately beneath the glacier sole. Water pressures of sufficient magnitude to allow deformation under considerable ice thicknesses are known to exist beneath many glaciers.

The stress at the glacier sole will attenuate at depth within the underlying sediments and ideally an expression is required for the maximum depth at which the basal shear stress would still be capable of deforming subglacial sediments. Unfortunately, such an expression cannot yet be derived.

The foregoing discussion suggests several generalizations. First, deformation of subglacial sediments as a result of shear stress imposed by overriding glacier ice only occurs when subglacial sediments are unfrozen. When the ice is thick, deformation is less likely and considerable pore-water pressures are required to produce it. Where the ice is very thin, deformation in unfrozen sediments is much more likely to occur, even when pore-water pressures are low. Unfrozen sediments beneath the extreme margin of a glacier are almost sure to be deformed.

However, it is commonly found that even relatively weak sediments lying beneath

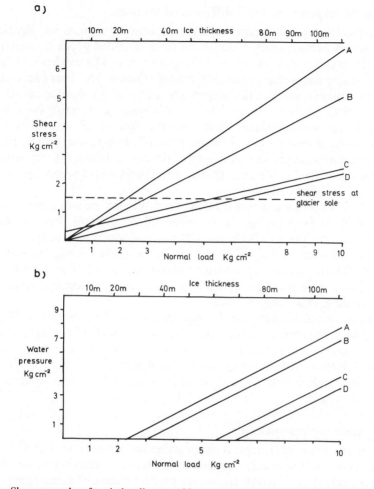

FIGURE 6a. Shear strengths of typical sediments with zero pore-water pressure. (A, gravel, $\phi = 34°$; B, loosely packed sand, $\phi = 27°$; C, clay, $C = 0·3$ bars, $\phi = 12°$; D, clay, $C = 0$, $\phi = 13°$) Above the dashed line, representing shear stress at the glacier sole, there will be no deformation.
b. The pore-water pressures required to allow deformation in sediments A, B, C and D under different ice thicknesses

presumed lodgement tills of Pleistocene age are undeformed. There are several ways in which this might occur. If the sediments beneath the advancing margin of a glacier are frozen (zone D) they would remain undeformed but, on subsequent thawing of these sediments under greater ice thicknesses (zones A, B or C), this overburden might be sufficiently great to prevent deformation. If the sediments beneath the advancing margin of a glacier are unfrozen, they will tend to be deformed. However, these deformed horizons might then be eroded beneath greater ice thicknesses, while subsequent deposition of a lodgement till upon lower undeformed horizons might bury them to a sufficient depth beneath the glacier sole to escape deformation on ice thinning and retreat. Finally, of course, the till might not be lodgement till.

Deformation in response to local differential stresses

Such deformation may occur beneath moving ice or beneath stagnant ice. Beneath moving ice it largely occurs because of irregularities on the glacier bed which concentrate stresses higher than the average value of stress at the glacier bed. The existence of lower stress values in intervening zones thus causes flow towards these zones. This phenomenon often occurs beneath relatively small ice thicknesses where cavities can develop on the lee sides of obstructions to flow. Many streamlined forms in lodgement till result from this process. Continuously fluted morainic ridges form from the flow of till into cavities on the lee sides of large boulders embedded in lodgement till (G. Hoppe and Schytt, 1953). Under greater ice thicknesses, the pressure in subglacial till may be inadequate to maintain the ice roof on the lee side of the boulder, and shorter drumlinoid ridges might thus form in these positions (R. W. Galloway, 1956).

Till ridges may also form where till flows from beneath ice into a completely unconfined situation. R. J. Price (1969) has described till ridges from Breiðamerkurjökull in south-east Iceland which lie parallel to the ice front, and which form at the ice margin as a result of the flow of water-soaked till from beneath the glacier. Along the eastern margin of the same glacier, till has also flowed from beneath the ice into a series of crevasses that intersect the frontal margin and lie normal to it. On ice retreat these give rise to a series of ridges at 90° to those described by Price.

Where large ice masses have largely stagnated, it has been suggested (Hoppe, 1952; A. Stalker, 1960) on the basis of Pleistocene evidence, that till might be squeezed into relict crevasses or beyond the margins of the masses producing parallel, annular or more complex till ridges. Such large-scale features can only be produced if there is a considerable thickness of unfrozen subglacial till. Thus these features are likely to develop on a large scale beneath temperate rather than cold ice.

Non-steady state considerations

The preceding discussion of thermal régimes makes the assumption that there is a steady thermal state. However, this is only an approximation. It is extremely unlikely that a glacier ever reaches a steady state. Thermal régime is a partial function of surface temperature and mass balance, and both may change strikingly as a result of climatic change. In large ice sheets, the time taken for thermal régime to adjust itself to climate is probably as long as or longer than many large-scale climatic cycles which may affect the glacier. In small valley glaciers, even annual changes in net budget and temperature may eventually affect the thermal régime, so that the latter is continually changing in response to annual or longer-term fluctuations. Thus it may be that processes which depend upon continual changes or oscillations of the boundaries of thermal zones are quite important, and do not necessarily rely on special pleading.

This type of variation might lead to alternations between processes discussed above. Alternatively, specific features might be produced by such alternations. For instance, oscillation of the B/C boundary could produce a series of thrust planes, or the type of incorporation by basal freezing envisaged by Weertman (1961). Phases of advance and retreat may also produce different phenomena. During advance, permafrost terrain may be overridden with little disturbance of frozen ground whereas, if retreat is relatively rapid, permafrost may not be able to establish itself sufficiently rapidly to retain a hold beneath the glacier margin. A temperate glacier transporting debris immediately above its bed is likely to produce lodgement till during both advance and retreat, but a cold glacier with debris transported at a relatively high level would tend to override this debris during advance, producing a subglacial till, but would release debris as supraglacial till during retreat.

It is also possible that two glaciers of similar size and under similar climatic conditions, but with different histories, might exhibit quite different thermal régimes. For instance, a glacier that has grown from a small size as a cold period develops might remain entirely cold, its sole frozen to the bed, whereas an existing large glacier whose sole was at the melting point might retain this condition during a subsequent cold phase because of the existence of heat derived from sliding. In the first case there is no heat derived from sliding and

$$\frac{(Tb - Ts)}{g^*}C > H_{\text{geo}}.$$

In the second case it may be that

$$\frac{(Tb - Ts)}{h^*}C \leqslant S + H_{\text{geo}} (+ Mw.L).$$

THE ROLE OF SUPRAGLACIALLY-DERIVED DEBRIS

The speculations presented above refer only to subglacial erosion, the incorporation and transport of the products of subglacial erosion, and the subglacial or supraglacial deposition of such material. The theory asserts overriding control by thermal régime. However, debris may also fall on to the glacier surface from nunataks, cirque headwalls and valley sides. This is not connected with thermal régime and does not apply to ice caps unpierced by nunataks. If such debris is acquired in the accumulation zone it becomes englacial; if it is acquired in the ablation zone it remains supraglacial. Ice-cored moraines in many glaciers, especially valley glaciers, largely occur because of a covering by supraglacially derived debris which is finally deposited at the ice margin as a sheet or as morainic ridges. In a valley glacier melting at its base, debris derived from a valley headwall may pass through the ice to be deposited as subglacial lodgement till.

REFERENCES

ANDERSON, J. L. and J. L. SOLLID (1971) 'Glacial chronology and glacial geomorphology in the marginal zones of the glaciers Midtdalsbreen and Nigardsbreen, South Norway', *Norsk. geogr. Tidsskr.* 25, 1–38

BIRD, J. B. (1959) 'Recent contributions to the physiography of northern Canada', *Z. Geomorph.* NF 3, 151–74

BOULTON, G. S. (1970) 'On the origin and transport of englacial debris in Svalbard glaciers', *J. Glaciol.* 9, 231–45

BOULTON, G. S. (1971) in *Till. A symposium* (ed. R. P. GOLDTHWAIT)

CAROL, H. (1947) 'The formation of *roches moutonnées*', *J. Glaciol.* 1, 57–9

EMBLETON, C. and C. A. M. KING (1968) *Glacial and periglacial geomorphology*

GALLOWAY, R. W. (1956) 'The structure of moraines in Lyngsdalen, north Norway', *J. Glaciol.* 2, 730–3

GARWOOD, E. J. and J. W. GREGORY (1898) 'Contributions to the glacial geology of Spitsbergen', *Q. J. geol. Soc. Lond.* 54, 197–227

GLEN, J. W., J. J. DONNER and R. G. WEST (1957) 'On the mechanisms by which stones in till become orientated', *Am. J. Sci.* 255, 194–205

GOLDTHWAIT, R. P. (1960) 'Study of ice cliff in Nunatarssuaq, Greenland', *Tech. Rep. Snow Ice Permafrost Res. Establ.* 39, 1–103

HOPPE, G. (1952) 'Hummocky moraine regions with special reference to the interior of Norrbotten', *Geogr. Annlr* 34, 1–71

HOPPE, G. and V. SCHYTT (1953) 'Some observations on fluted moraine surfaces', *Geogr. Annlr* 35, 105–15

JELLINEK, H. H. G. (1959) 'Adhesive properties of ice', *J. Colloid Sci.* 14, 268–80

KAMB, B. and E. LACHAPELLE (1964) 'Direct observation of the mechanism of glacier sliding over bedrock', *J. Glaciol.* 5, 159–72

LINTON, D. L. (1955) 'The problem of tors', *Geogrl J.* 121, 478–87

LOEWE, F. (1966) 'The temperature of the Sukkertoppen ice cap', *J. Glaciol.* 6, 179

MATHEWS, W. H. (1964) 'Water pressure under a glacier', *J. Glaciol.* 5, 235–40

McCALL, J. G. (1960) 'The flow characteristics of a cirque glacier and their effect on glacial structure and cirque formation' in 'Norwegian cirque glaciers' (ed. W. V. LEWIS), *R. geogr. Soc. Res. Ser.* 4, 39–62

PRICE, R. J. (1969) 'Moraines, sandar, kames and eskers near Breiðamerkurjökull, Iceland', *Trans. Inst. Br. Geogr.* 46, 17–43

ROBIN, G. DE Q. (1955) 'Ice movement and temperature distribution in glaciers and ice sheets', *J. Glaciol.* 2, 523–32

SCHYTT, V. (1969) 'Some comments on glacier surges in eastern Svalbard', *Can. J. Earth Sci.* 6, 867–73

STALKER, A. (1960) 'Ice-pressed drift forms and associated deposits in Alberta', *Bull. geol. Surv. Can.* 57, 1–38

SVERDRUP, H. V. (1935) 'Scientific results of the Norwegian-Swedish Spitsbergen Expedition in 1934; part III', *Geogr. Annlr* 17, 53–88

TABOR, D. and J. C. F. WALKER (1970) 'Creep and friction of ice', *Nature, Lond.* 228, 137–9

TANNER, V. (1938) 'Die Oberflächengestaltung Finnlands', *Bidr. Känn. Finl. Nat. Folk* (Helsingfors), 86

VIVIAN, R. (1970) 'Hydrologie et érosion sous-glaciaires', *Rev. Géogr. alp.* 58, 241–64

WEERTMAN, J. (1961) 'Mechanism for the formation of inner moraines found near the edge of cold ice caps and ice sheets', *J. Glaciol.* 3, 965–78

WEERTMAN, (1966) 'Effect of a basal water layer on the dimensions of ice sheets' *J. Glaciol.* 6, 191–207

RÉSUMÉ. *Le rôle du régime thermal dans la sédimentation glaciaire.* Le régime thermal d'un glacier, déterminé par le climat, le bilan de masse, et l'épaisseur de la glace, peut être employé afin de définir quatre états frontières à la plante d'un glacier. Ceux sont: une zone (A) de nette fusion basale, une zone (B) dans laquelle se trouve une balance entre la fusion et le regel, une zone (C) dans laquelle suffisante eau de fonte gèle à la plante du glacier afin de la soutenir au point de fusion, et une zone (D) dans laquelle la plante du glacier est au dessous du point de fusion. Dans la zone A, le glacier glisse au dessus du lit et l'érosion se produit par le moyen d'abrasion et d'écrasement. Les produits d'érosion sont transportés dans une couche mince de fusion et regel, pas plus que 0,5–1,0 m d'épaisseur, et dans la suite ils sont déposés comme argile de logement (« lodgement till ») dans les zones où le retardement de friction contre le lit est grand. Les matériaux sous-glaciaires sont facilement déformés à cause des hautes poussées de l'eau sous-glaciaires. Dans la zone B, les processus sont essentiellement pareils, mais les vitesses d'érosion et de dépôt peuvent bien différer à cause des plus basses poussées de l'eau sous-glaciaires. Dans la zone C, le glacier glisse encore au dessus du lit, mais l'arrachement du matériel sous-glaciaire a lieu en outre d'abrasion et d'écrasement. Les produits d'érosion peuvent être transportées aux hauts niveaux dans la glace à cause du net gel d'eau à la plante du glacier. En raison du haut niveau de transport, le dépôt subséquent tend à prendre la forme d'argile de fluence (« flow till ») et argile de fusion (« melt-out till ») plutôt qu'argile de logement. Dans la zone D, le glacier se colle au lit, il n'y en pas a de glissement basal et le mouvement ne se produit que par courant intérieur. L'abrasion et l'écrasement sont des processus relativement sans importance, mais l'arrachement de très grands blocs erratiques peut bien arriver. Il y a encore peu de dépôt d'argile de logement sous-glaciaire, car l'éboulis transporté au dessus de la plante du glacier est en grande partie déposé comme argile de fusion et argile de fluence. Déformation des sédiments sous-glaciaires gelés n'arrive pas. Dans tous ces cas, les variables majeures qui influent sur le processus sont l'âpreté du lit, le débit d'eau sous-glaciaire, et la perméabilité des strates sous-glaciaires.

FIG. 1. Des façons dans lesquelles des zones thermales différentes peuvent être combinées. Chaque zone a des processus caractéristiques à elle, d'érosion, de transport, de dépôt, et de déformation de sédiment. Les différentes manières dans lesquelles les zones sont combinées influent aussi sur les processus à un point quelconque

FIG. 2. L'action de la perméabilite (K) des gisements sous-glaciaires sur l'épaisseur d'une couche basale d'eau. L'épaisseur de la couche d'eau règle la poussée effective entre la glace et le lit, et ainsi influe sur l'érosion et le dépôt

FIG. 3. Lignes de flux plastique au dessus du lit irrégulier d'un glacier auquel on n'en trouve pas de glissement au contact entre la glace et le lit

FIG. 4. L'arrachement sous-glaciaire des grands blocs erratiques dans la zone D où le sol sous-glaciaire gelé en permanence est mince

FIG. 5a. La couche mince de fusion et regel, chargée d'éboulis, à la base de la glace tempérée. Sous poussée fonte (P) a lieu sur les flancs des obstacles en amont du glacier, et regel (R) de l'eau de fonte résultante (indiqué par une flèche) a lieu sur le flanc en aval du glacier.

b. La nette acquisition des glaciers dans la zone C de glace basale chargée d'éboulis. Des suites d'éboulis épaisses peuvent ainsi accumuler.

c. Le changement de la zone C (ou B) à la zone D, qui entraîne la cessation de glissement au lit. La compression résultante occasionne des faces de chevauchement le long desquelles glace basale peut être transportée à un niveau plus haut

FIG. 6a. Les forces de cisaillement des sédiments typiques aux nulles poussées de l'eau dans les pores. (A, gravier, $\phi = 34°$; B, sable peu tassé, $\phi = 27°$; C, argile, C = 0·3 bars, $\phi = 12°$; D, argile, C = 0, $\phi = 13°$). Au dessus de la ligne des traits, qui représente l'effort de cisaillement à la plante du glacier, il n'y en aura pas de déformation.

    b. Les poussées de l'eau exigées dans les pores afin de permettre la déformation du sédiments A, B, C et D, sous des épaisseurs de glace diverses

---

ZUSAMMENFASSUNG. *Die Rolle des Thermalregimes in der Glazialsedimentation.* Das Thermalrégime eines Gletschers, bestimmt durch Klima, Massenbilanz und Dicke des Eises, kann dazu gebraucht werden, um vier Grenzzustände an der Gletschersohle zu definieren. Diese sind: eine Zone (A) der netto Grundschmelzung, eine Zone (B), in der ein Gleichgewicht zwischen Abschmelzung und Gefrieren besteht, eine Zone (C), in der genügend Schmelzwasser an die Gletschersohle friert, um sie am Schmelzpunkt zu halten, und eine Zone (D), in der die Gletschersohle unter dem Schmelzpunkt liegt. In der Zone A gleitet der Gletscher über sein Bett und Erosion findet durch Abreibung und Zermalmung statt. Die Erosionsprodukte werden in einer dünnen Regelationsgrundschicht, die nicht mehr als 0,5–1,0 m dick ist, transportiert, und werden danach als Festmergel (‚lodgement till‘) in Zonen abgelagert, in denen die Reibungsverzögerung gegen die Schicht stark ist. Subglaziale Materialien werden wegen hoher subglazialer Wasserdrücke leicht verformt. In Zone B sind die Vorgänge im Grunde ähnlich, aber Grad und Geschwindigkeit der Erosion und Ablagerung können sehr wohl wegen der niedrigen subglazialen Wasserdrücke unterschiedlich sein. In Zone C gleitet der Gletscher noch über sein Bett, aber zusätzlich zu dem Losreissen des subglazialen Materials geschieht auch Abreibung und Zermalmung. Die Erosionsprodukte können durch das netto Gefrieren des Wassers an der Gletschersohle bis in die hohen Stufen des Eises hinaufgetragen werden. Wegen der hohen Transportstufe neigt die folgende Ablagerung dazu, eher die Form von Fliess- und Schmelzmergel als von Festmergel anzunehmen. In Zone D haftet der Gletscher an seinem Bett, es gibt kein Grundsohlengleiten und die Bewegung findet allein durch inneren Fluss statt. Abreibung und Zermalmung sind verhältnismässig unwichtige Vorgänge, aber Losreissen von sehr grossen erratischen Blöcken kann sehr wohl vorkommen. Wieder gibt es wenig Ablagerung des subglazialen Festmergel, weil der Schutt, der über der Gletschersohle getragen wird, hauptsächlich als Schmelz- und Fliessmergel abgelagert wird. Verforming der gefrorenen subglazialen Sedimente kommt nicht vor. In alle diesen Fällen sind die Rauhheit des Bettes, der Ausfluss subglazialen Wassers, und die Durchlässigkeit der subglazialen Schichten die Hauptveränderlichen, die sich auf den Vorgangauswirken.

ABB. 1. Mögliche Verbindungsweisen verschiedener Thermalzonen. Jede Zone hat ihre eigenen charakteristischen Erosions-, Transports-, Ablagerungs- und Sedimentverformungsvorgänge. Die Verbindungsanordnungen der Zonen beeinflussen auch an irgend einem Punkt die Vorgänge

ABB. 2. Die Wirkung der Durchlässigkeit (K) subglazialer Schichten auf der Dicke einer Grundwasserlage. Die Dicke der Wasserlage regelt den wirksamen Druck zwischen Eis und Bett und beeinflusst dadurch Erosion und Ablagerung

ABB. 3. Plastische Fliesslinien über einem unregelmässigen Gletscherbett, in dem es kein Gleiten an der Eis-/Bett-berührung gibt

ABB. 4. Das subglaziale Losreissen grosser erratischer Blöcken in Zone D, wo der subglaziale Dauerfrost dünn ist

ABB. 5a. Die dünne schuttbelastete Regelationsschicht an der Unterseite von gemässigtem Eis. Druckschmelzung (P) findet an der gletscheraufwärtigen Hindernisseite statt, und Regelation (R) des entstehenden Schmelzwassers (durch einen Pfeil gekennzeichnet) findet an der abwärts Gletscherseite statt.

    b. Der netto Erwerb von schuttgeladenem Grundeis der Gletscher in Zone C. Dicke Schruttreihen können sich so ansammeln.

    c. Die Wandlung von Zone C (oder B) zur Zone D, welche das Einstellen von Bettgleiten mit sich bringt. Die entstehende Verdichtung verursacht Überscheibungsflächen, auf denen Grundeis auf eine höhere Stufe entlang transportiert werden kann

ABB. 6a. Scherstärken typischer Sedimente mit Porenwasserdruck auf dem Nullpunkt. (A, Kies, $\phi = 34°$; B, leicht gepäckter Sand, $\phi = 27°$; C, Ton, C = 0·3 bars, $\phi = 12°$; D, Ton, C = 0, $\phi = 13°$). Über der Strichlinie, die die Scherspannung an der Gletschersohle bezeichnet, wird es keine Verformung geben.

    b. Die nötigen Porenwasserdrücke, um Verformung der Sedimente A, B, C und D unter verschiedenen Eisdicken zu ermöglichen

The wear of sandstone by cold, sliding ice

R. HOPE, H. LISTER and R. WHITEHOUSE

Department of Geography, University of Newcastle-upon-Tyne

Revised MS received 9 November 1971

ABSTRACT. Rock wear by sliding ice is considered a function of friction. Laboratory experiments, briefly described, show friction increasing with load and with fall in temperature but decreasing generally with speed and with duration of sliding. This seems to be caused by the formation, at the ice sliding surface, of a thin layer of small deformation crystals with *c* axis normal to the surface; ice so orientated has a very low coefficient of friction. Increase in load or speed of sliding causes a jerky motion. This stick-slip motion frequently leaves ice adhering to the rock, producing an ice-to-ice sliding surface. The sliding ice shows lines of rock particles analogous to dirt bands in glaciers. Production of rock debris by sliding ice is small, but shatter of pebbles is readily observed after only seven cycles of freeze-thaw.

To measure rock wear by ice sliding over it, a 7-cm diameter cylinder of medium-grained silica sandstone was rotated about its vertical axis with weights on top to give an adjustable loading on the circular rock face. The ice was restricted in turning by a spring balance to measure the torque produced by the friction. The apparatus was built to fit inside a refrigerator with coarse temperature control.

Examination of the sliding surfaces revealed rock wear at the centre of the rock face but some polishing in patches nearer the edge. The speed of sliding varied from zero at the centre of the rock to more than 100 m per day at the periphery. Measurement of profiles across the circle of rock generally corroborated the results of a previous experiment in which the wear produced by ice sliding in a sinusoidal motion over rectangular rock surfaces was measured (H. Lister *et al.*, 1968).

The rotation experiment was repeated with loads of 0·5, 1 and 2 kg cm^{-2} and temperature maintained at $-12°C \pm 1°$. Periodic examination revealed patches of rock flour with ice crystals and small pits near the central, deeper pit. Figure 1 shows a typical example. The experiments indicated a decrease in the rate of rock wear with (i) time of sliding, (ii) increase in sliding speed, and (iii) decrease in particle size of the silica sandstone used.

To avoid a possible drilling action through the rock cylinder, a series of annuli was cut, the largest 8 mm wide at 4 cm radius. Precise measurement of rock wear required repeated removal of the rock from the refrigerator and thus subjected it to a form of freeze-thaw erosion. To preclude this, it was decided to accept the premise that wear is directly indicated by values of friction at the sliding surface. The spring balance measuring torque caused by the frictional force was fitted over a potentiometer with linear range, measured outside the refrigerator by a recording galvanometer. The ice was prepared as an agglomerate of broken crystals with super-cooled water, added slowly to encourage retention of random orientation.

Loads up to 10 kg cm^{-2} were applied to represent an ice thickness up to 110 m. A glacier of that depth provides a thermal insulation such that the sliding surface could be maintained at or very near to melting point, particularly in sub-polar regions with positive

FIGURE 1. Wear of medium-grained silica sandstone by ice rotating over it produces maximum erosion at the slower speeds. Products of wear can be seen in the central pit. The 7 cm diameter rock circle was a plane surface before the experiment

temperatures in summer and hence abundant seasonal meltwater. Since the experiment was aimed at measuring rock wear by cold, sliding ice, the loads applied were generally in the equivalent thickness range of 10 to 100 m.

Sublimation of the ice compelled replacement every two weeks. The distance travelled at observed glacier speeds is exceedingly small in this short time, so the sliding speeds used were in the range from 1 to 50 m day^{-1}.

Friction and load

The slope of each curve in Figure 2a is the coefficient of friction (μ) which changes very little with applied load, but it can be seen in Figure 2b that though small, there is a constant linear increase of μ with the logarithm of pressure on the sliding surface. This increase of μ with applied load is the same here for sandstone and for marble, though the value of friction decreases with speed of sliding for sandstone but increases with speed of sliding over marble. D. Tabor and J. F. C. Walker (1970) found that, for ice sliding on polished granite but with much higher loads than used here, μ increased with speed to 1 m day^{-1} but decreased with further increase in sliding speed.

Friction and temperature

The coefficient of friction is low and somewhat erratic at temperatures near to 0°C but

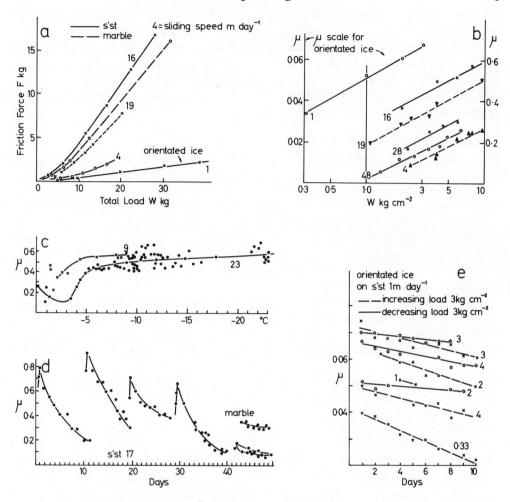

FIGURE 2. Friction at a sliding ice/rock surface as a function of load, temperature and time. The ice crystals are in random orientation except where stated as orientated ice, when the *c* axis was normal to the sliding surface. The non-dimensional figures indicate the sliding speed in m day^{-1} except in (e) where the figure is the load in kg cm^{-2}. Temperature was $-7°$C except in (c)

increases until at $-5°$C there is a very low rate of increase with further fall in temperature (Fig. 2c).

The real area of sliding contact, particularly in the case of a sandstone, must be less than the apparent area because of imperfections in the machine-cut surface and the space round each grain; the latter is more evident when the surface is worn as shown in Figure 1. If the granular surface provides sliding contact over half the area, the stress is doubled. Here the true contact area is probably much less than half and the local stress is thus considerably increased, but the loads applied were too small to cause pressure melting except perhaps at localized asperities on initial contact (Fig. 2d) and over the first few degrees of freezing as shown in Figure 2c.

Heat calculated from the friction values was very low and, for these models in the refrigerated environment, could maintain only a very low temperature gradient.

P. Barnes and Tabor (1966) measured ice hardness with loading time of spherical and conical identors and found that hardness increased exponentially as the temperature decreased. For a constant strain rate, the applied stress from Glen's creep law behaved similarly. Friction, a function of hardness, here increased linearly as temperature decreased but the scatter precludes any definite conclusion.

Friction and time of sliding

Figure 2d is the record of friction at $-7°C$ with four successive ice blocks on the same annulus of medium-grained sandstone sliding at 17 m day^{-1} with apparent loading of 2 kg cm^{-2}. Pressure melting as the surfaces came together may have caused the initial value of friction to be lower than the peak values found after a few metres of sliding. Subsequent values suggest a decreasing rate of fall in friction with duration of sliding and it is unfortunate that sublimation of the ice prevented continued sliding with the same ice block until steady values of friction were recorded. Ice on marble gave a steadier coefficient of friction in a much shorter time. The marble wore more readily, giving fine grains that apparently lubricated the sliding.

The change of friction with time was partly responsible for the scatter of values shown in Figure 2c. It was thought at first that the decrease in friction of the sliding over sandstone was the result of wear but a new ice block, sliding on the same rock, raised the friction value, suggesting that a change in the ice surface during sliding was the dominant cause. Repeating the experiment with the same rock and also with finer grained sandstone which had been more carefully smoothed gave similar results, but in all cases the spread of values was an embarrassment and it was difficult to achieve repeatable values with apparently identical conditions.

The sliding surface

Thin sections of the ice (Fig. 3), cut after sliding, revealed a base of coarse crystals with c axis in various directions but there were never enough crystals to permit a firm conclusion that these directions were completely random. More interesting was the sliding surface of the ice which showed, at the base of the initial crystal mass, a thin layer of much smaller crystals, many orientated almost normal to the sliding surface.

Observations of the ice/rock contact surface beneath glaciers have generally shown a dark layer of ice to which B. Kamb and E. LaChapelle (1964) gave the generic term 'regelation layer'. Barnes and G. de Q. Robin (1966) suggest regelation-deformation owing to shearing of the ice over bedrock. The sole of cold glaciers is also dark, but contains much more rock debris than in the case of temperate ice. G. S. Boulton (1970) describes bands near to the ice/rock surface as having a preferred crystal orientation. The rock particles with small crystals in Figure 3 represent a micro dirt band.

Friction with orientated crystals

To check on the importance of crystal orientation near the ice-rock interface, blocks of ice were prepared in a manner devised by Barnes (1968). Boiled, deionised water, in a reservoir connected from below to a heated expansion chamber, was frozen at a controlled rate progressing downwards from the water surface, where all the heat exchange took place.

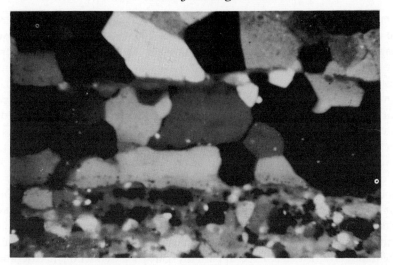

FIGURE 3. Thin sections of ice at the sliding surface. The large crystals at the top of the photograph are the initial ice. The bottom third is a dirt band 1 mm thick with rock flour showing as a grey area round the small ice crystals, the products of recrystallization by the shearing

The crystals were thus orientated with c axis normal to the ice surface. The block of ice was rotated over a 19·3 cm² rock annulus of the same medium-grained sandstone as used previously but the rock had been cleaned in an ultra-sonic vibrator. An improved apparatus permitted lower sliding speeds with the friction force measured by a strain-gauge transducer, the output from which was shown as a trace on a recording galvanometer. Each regression line shown in Figure 2e represents the decrease in the coefficient of friction over 10 days of sliding at $-7°C$ and 1 m day⁻¹ with the applied load shown. The low values of friction (reduced by a factor of 10) are remarkable but are in accordance with the findings of Tabor and Walker (1970) who, on a single crystal with c axis parallel to the direction of sliding, found recrystallization in a thin layer of ice with c axis normal to the interface. 'This orientation is favourable to easy shear tangential to the surface . . . Creep in this direction is about 100 times easier than when ice is in other orientations' (Tabor and Walker, ibid.). Thin sections of the ice showed the initial crystals with a thin layer of smaller crystals at the sliding surface, both with c axis normal to the surface. Such orientation limits creep into the rock voids and is a further possible reason for reduction of friction with time of sliding.

Change of friction with time of sliding is not constant for a specific load but is so shown in Figure 2e because only a linear regression could be calculated through the observed values. The correlation coefficient for each regression decreased from 0·983 at the lowest loading to 0·708 at the highest loading of 4 kg cm⁻². Creep, caused by the increase in load, seems to be the factor reducing the correlation. The slope of the regression lines decreases as the load was increased but mean values of friction after 3 or 4 days sliding (the earliest time at which repeatable values had been found in previous experiments) showed the same proportional increase with the logarithm of the load (Fig. 2b) up to 3 kg cm⁻². Rock wear was scarcely perceptible up to this load but at 4 kg cm⁻² became visible in the form of a few grains of rock flour with two or three grains from the parent rock, some facetted, most probably by the initial polishing. The reduced slope of the regression lines with time of sliding, even when loads were decreased, was thought to be caused by this wear changing the true contact surface, the dominant asperities having been removed.

If we assume that the load of 4 kg cm^{-2} was effective over half the sliding area, the strain over the time of sliding across one intergranular space or pit (which was rarely more than 1 mm) is given by the creep law as 10^{-5}. With the crystal orientation normal to the sliding surface, the strain is considerably less than this; hence, for the small dimensions involved here, the creep was insufficient to fill the cavities between grains. Thus the fall in friction between loads of 3 and 4 kg cm^{-2} was not caused by creep completing the contact area and reducing the applied stress.

The orientated ice curtails creep in the vertical direction which severely limits the area of contact; local stress can thus be considerable. The reduction of melting point is only 0·007°C bar^{-1}, so that a loading of 1000 bars is necessary here for pressure melting. This requires that, with an apparent stress of 4 kg cm^{-2}, the contact area is approximately 1/250 of the apparent area. This is only possible at localized asperities but, since pressure melting of randomly orientated ice apparently caused a reduction in friction down to −5°C (as shown in Figure 2c), it seems to be the only possible explanation for the sudden reduction in friction between loads of 3 and 4 kg cm^{-2} as shown in Figure 2e. The rock wear evident at these loads was probably also associated with pressure melting.

Since pressure melting was present only at local points and the ice did not creep over the complete surface, J. Weertman's (1957) theory of basal sliding as a function of bed roughness did not apply. However, it is interesting to find that the Weertman velocity, dependent upon creep over the roughness elements, with adjustment for the lower temperatures, was only of the order of millimetres per year, i.e. approximately zero, which is almost equivalent to the ice being frozen to the bed.

Stick-slip motion

The few continuous records of ice movement near bedrock show large variations in rates of flow, e.g., R. P. Goldthwait (1969) recorded a jerky motion for an Alaskan glacier, but in cold glaciers where the ice is frozen to the rock (e.g. parts of Greenland), shearing takes place in the ice.

In experiments, increased loading of randomly orientated ice, sliding at temperatures appropriate to cold ice, generally showed a jerky motion with the ice adhering to the rock until the static friction was overcome and slip occurred. Static friction increased rapidly with decrease in temperature and, during the 'stick' phase of this stick-slip motion, the ice block adhered to the rock, separation from which revealed smoothed patches of ice so that sliding was no longer over a true ice/rock contact (Fig. 4). This cold welding with subsequent shearing of asperities is a common form of friction in engineering materials (F. P. Bowden and Tabor, 1950). Ice crystals orientated with c axis parallel to the rock surface could much more readily creep into the voids in the rock surface and, though the creep rate was too slow for this to be possible in the continuously sliding ice, it was much more possible during the 'stick' phase in stick-slip motion. Simulation of this motion by sliding sandstone to and fro under weighted ice blocks produced ice accretion on the rock proportional to the particle size (and hence to the void space) as shown in Figure 5.

Summarizing these results on cold ice, the amount of wear observed in these sliding experiments was small and the early formation of a basal deformation layer of orientated

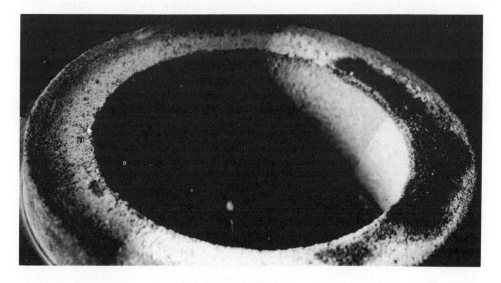

FIGURE 4. Rock annulus (7 cm diameter) after sliding under ice at $-6°C$ with a load of 2 kg cm^{-2} and a stick-slip motion. The dark patches are ice adhering to the rock

FIGURE 5. Sandstones after sliding to and fro (stick-slip motion) under ice. The amount of ice adhering is proportional to the particle size of the different sandstones

FIGURE 6. Vertical section (15 cm wide) of four layers of gravel alternating with four layers of very fine sand, after seven cycles of freeze-thaw. Relative movement upwards, merging of the two upper-layers and shatter of the coarser particles is most marked

ice, causing a marked reduction in friction and hence in rock wear, soon curtailed the erosion. Irregularity in sliding speed, particularly if developed to a stick–slip motion, can quickly produce an ice-to-ice contact and hence little rock wear.

Freeze-thaw

The problem remains of how so much debris is made available under sub-polar glaciers. From extensive observations in Spitsbergen, Boulton (1970) favours a form of Weertman's basal freezing hypothesis: '... Unconsolidated subglacial sediments would thus tend to freeze to the glacier sole, and slip would occur at the junction between the frozen and underlying unfrozen sediments'. This implies a fluctuation about freezing point for part of the sediment layer. This fluctuation is also inherent in the change from temperate to polar condition of the ice-rock contact as ice moves from the interior or accumulation zone to the edge of a cold ice cap.

To measure the effectiveness of such temperature fluctuations, two beds of sand with gravel were prepared with the gravel, sized and in four discrete layers, alternating with a layer of very fine sand. Water was added until the whole was just wetted. Each matrix was held in a container 15 cm in diameter, well insulated round the sides and base to limit heat exchange to the vertical direction. One container was taken through seven and one through fifteen cycles of freeze-thaw, each of 72 hours with a freezing temperature of $-6°C$. The containers were then jacketted by ice to permit the specimens to be held rigidly while they were cut along a diameter. Figure 6 shows one specimen, after seven cycles, with gravel layers buckled and the two upper layers merged. Shattering of the bigger gravel, especially the sedimentary pebbles, is quite marked. This was also evident from the top view. A freezing process thus seems to be more effective than sliding ice in the production of rock debris.

Furthermore, the typical form of a glaciated valley suggests that sub-aerial erosion of rock above the ice surface must be more effective than glacial erosion along the valley side.

Although there are rare examples of glaciated valleys with apparently overhanging valley walls (V. Haynes, personal commun.), this is very unusual and contrary to the hypothesis that glaciers do not undercut the valley wall. It would seem that freeze-thaw is at least as effective as sliding ice in rock erosion.

CONCLUSION

Greater control is required in laboratory experiments such as those described above, and there is a need to repeat them over a wider range of rock types and speeds of sliding. Our knowledge of glacier erosion as succinctly summarized by C. Embleton and C. A. M. King (1968) is only modified by these experiments in three points: (i) Sliding friction and hence rock wear by cold ice is not a simple function of sliding speed; on hard rock, low speeds produce higher friction. (ii) Localized pressures can be very high indeed and sliding friction erratic down to $-5°$ or $-6°C$. Fluctuation through lower temperatures with high loads produces a stick-slip motion and an ice-to-ice sliding surface. (iii) A basal layer of deformation ice forms readily and reduces sliding friction and wear; hence, if loads are sufficient, rock wear is greatest in the initial period of sliding, decreasing rapidly with time. Re-sharpening of glacial features by local ice after regional recession is thus very possible.

DISCUSSION

G. S. BOULTON: Would Dr Lister comment on the extent to which some of his findings on wear by sliding ice are directly applicable to real glaciers? The coefficients of friction which he finds for ice in which c-axes are not preferentially oriented normal to the sliding surface seem to lie between 0·2 and 0·8. It has been shown both theoretically and by direct measurement that the average shear stress at the sole of a glacier is of the order of 0·5 to 1·5 kg cm^{-2}. This being so, and if Dr Lister's figures are applied directly, ice at the temperatures measured by Dr Lister would not slide over its bed under thicknesses of ice greater than between 7 m (using $\mu = 0·8$, shear stress = 0·5 kg cm^{-2}) and 75 m (using $\mu = 0·2$, shear stress = 1·5 kg cm^{-2}). Thus, the sliding of cold ice over its bed seems unlikely to be a significant agent of erosion, unless basal ice is highly oriented.

Has Dr Lister determined the relationship between increased speed of sliding and reduced wear other than by using rotating cylinders of rock? I am not convinced that the reduced linear velocity near the centre of the cylinder is entirely responsible for the wear produced there. The rotational torque transmitted by the ice to the component particles of the sandstone near the centre of rotation is much greater than it is farther from the centre, and may effectively wrench particles from the sandstone.

H. LISTER: Any physical model of a large-scale natural phenomenon poses many problems, particularly of scale. Sliding speeds used in the models shown were high; sliding was artificially initiated at various speeds to permit measurement of friction and rock wear. The data presented were recorded from the model, using a rock annulus to minimize the range of torque. Difference in torque across a sand grain cannot be great, so it is not likely to wrench particles from the rock. However, the annulus avoided the centre, where the sandstone showed considerable wear. I agree with the approximate ice thicknesses given by Dr Boulton to provide Nye's average shear stress for glacier sliding, but here using friction values from the model. They represent a fair range of glacier thickness, under which rock wear seems small: it is yet smaller when crystal reorientation reduces friction. However, at low temperatures, with a slip-stick motion, ice adheres to the rock and shearing takes place in the ice; such ice-to-ice sliding certainly produces little wear.

ACKNOWLEDGEMENT

We gratefully acknowledge support from the Natural Environment Research Council for part of this work.

REFERENCES

BARNES, P. and D. TABOR (1966) 'Plastic flow and pressure melting in the deformation of ice, I,' *Nature, Lond.*, 210–5039, 878–83

BARNES, P. and G. de Q. ROBIN (1966) 'Implications for glaciology', *Nature, Lond.* 210–5039, 878–83

BARNES, P. (1968) 'Mechanisms of deformation in ice. I', (Unpubl. Ph.D. thesis, Cambridge University)

BOULTON, G. S. (1970) 'On the origin and transport of englacial debris in Svalbard glaciers', *J. Glaciol.* 9, 213–29

BOWDEN, F. P. and D. TABOR (1950) *The friction and lubrication of solids* (Oxford Univ. Press)
EMBLETON, C. and C. A. M. KING (1968) *Glacial and periglacial geomorphology* (London)
GOLDTHWAIT, R. P. (1969) 'Continuous measurement of glacier movement in Alaska' (*Cambridge Glacier Hydrol. Conf.*, to be published)
KAMB, B. and E. LACHAPELLE (1964) 'Direct observation of the mechanism of glacier sliding over bedrock', *J. Glaciol.* 5, 159–72
LISTER, H., A. PENDLINGTON and J. CHORLTON (1967) 'Laboratory experiments on abrasion of sandstones by ice', *Int. Ass. scient. Hydrol.*, *Commn Snow Ice* (Bern), 98–106
TABOR, D. and J. F. C. WALKER (1970) 'Creep and friction of ice', *Nature, Lond.* 228–5267, 137
WEERTMAN, J. (1957) 'On the sliding of glaciers'. *J. Glaciol.* 3, 33–8

RÉSUMÉ. *L'usure du grès par la glace froide qui glisse.* On regarde l'usure du rocher par le glissement de la glace comme une fonction de la friction. Les expériences du laboratoire, en somme, révèlent que la friction augmente avec le taux de charge et avec une baisse de temperatur mais que la friction diminue, en général, avec la vitesse et avec la durée du glissement. Il semble que ce soit à cause de la formation de petits cristaux déformés, dont le *c* axe est perpendiculaire à la surface glissante; la glace, tellement orientée, a un coefficient de friction très bas. L'augmentation de taux de charge ou de la vitesse glissante fait naître un mouvement saccadé. Ce mouvement, « le stick-slip » laisse fréquemment quelque glace, qui adhère au rocher et qui produit une surface glissante de la glace sur la glace. Le glissement de la glace révèle les lignes de particules de rocher analogues aux bandes saletées du glacier. La production de détritus par la glace qui glisse est petite mais l'on n'observe qu'après sept cycles d'un cours de la gelée et du dégel un fracassement de cailloux.

FIG. 1. L'usure du grès silice de grain moyen par la glace qui pivote au-dessus, produit l'érosion la plus grande aux vitesses les plus lentes. On voit les productions de l'usure dans le trou central. Le cercle de rocher, qui avait une diamètre de 7 cm, était une surface plane avant l'expérience
FIG. 2. La friction à la surface entre la glace et le rocher en tant qu'une fonction du taux de charge, de la température et du temps. Les cristaux de glaces sont en orientation hasarde sauf ceux qu'on a affirmé d'être en glace orientée, quant l'axe *c* était perpendiculaire à la surface glissante. La figure est non-dimensions indique une vitesse du glissement en m jour⁻¹, sauf la figure 2e où l'on donne le taux de charge à kg cm⁻². La temperature était −7°C sauf la figure 2c
FIG. 3. Une section peu épaisse de la glace, à la surface glissante. Les cristaux grands en haut de la photographie sont la glace originaire. La partie la plus basse est une bande saletée, 1 mm d'épaisseur, avec la poudre du rocher révèle comment une zone grise autour des petits cristaux de glace, les produits de récristallisation par le cisaillement
FIG. 4. Le rocher annulaire qui a une diamètre de 7 cm après s'être glissé sous la glace à la temperature de −6°C et au taux de charge de 2 kg cm⁻² et au mouvement du « stick-slip ». Les pièces sombres sont quelque glace, qui adhère au rocher
FIG. 5. Les grès après s'être glissé d'un mouvement de va-et-vient (le mouvement du « stick-slip ») sous la glace. La glace qui adhère est proportionelle à la dimension des particules de grès différents
FIG. 6. Une coupe verticale, 15 cm de large, de quatre bandes du gravier qui alternent de quatre bandes de sable très fin, aprés sept cycles d'un cours de la gelée et du dégel. Un mouvement relatif vers le haut, la fonte des deux bandes supérieures et le fracassement des particules plus grossières sont des effects les plus évidents

ZUSAMMENFASSUNG. *Die Zerfressung des Sandsteins durch das kalte Eisgleiten.* Man hält die Zerfressung eines Felsen durch das Eisgleiten für eine Funktion der Reibung. Einige Versuche im Laboratorium zeigen, kurz zusammengefasst, dass die Reibung steigert, wenn die Last sich erhöht oder die Temperatur sich vermindert, doch im Allgemeinen abfällt, wenn die Geschwindigkeit und die Dauerzeit des Gleitens zunimmt. Die Ursache davon scheint auf die Gestaltung einer dünnen Schicht von kleinen Deformationskristallen auf der Gleitoberfläche des Eises hinzudeuten, deren *c* Achse rechtwinkelig auf dem Oberfläche ist; wenn das Eis so orientiert wird, hat es einen sehr niedrigen Reibungskoeffizient. Wird die Last oder die Gleitungsgeschwindigkeit erhöht, so entsteht ein plötzlicher Stoss. Diese ‚Stick-slipbewegung' lässt häufig etwas Eis hinter sich, das am Felsen klebt und eine Gleitoberfläche von Eis auf Eis produziert. Im gleitenden Eis sind dünne Schichten von Felspartikeln zu sehen, die Gletscherschmutzbändern ähnlich sind. Das Eisgleiten produziert wenig Schutt aber schon nach sieben Kreisläufen eines Frieren- und Tauenprozesses ist das Zerschmettern von Kieselsteinen leicht zu merken.

ABB. 1. Die Zerfressung des mittelgranulierten kieselartigen sandsteins durch das Eis, welches daüber rotiert, produziert die höchste Erosion mit den langsameren Geschwindigkeiten. Die produkte der Zerfressung sind in der mittleren Grube zu sehen. Der kreisförmige Fels, der einen Durchmesser von 7 cm hatte war eine Fläche vor dem Experiment
ABB. 2. Die Reibung an einer Oberfläche von gleitendem Eis und vom Felsen, als Funktion der Last, der Temperatur und der Zeit. Die Eiskristallen sind von zufälliger Orientierung, es sei denn dass man sie als orientierte Eiskristallen bestäligt, wo die *c* Achse rechtwinkelig auf der Gleitgeschwindigkeit in m Tag⁻¹ ausser im Abb. 2e wo die Zahl die Last in kg cm⁻² darstellt. Die Temperatur war −7°C ausser im Abb. 2c

ABB. 3. Ein dünner Abschnitt Eis an der Gleitoberfläche. Die grossen Kristallen oben am Photo sind das ursprüngliche Eis. Das untere Teil ist ein Schmutzband (mit einer Breite von 1 mm) mit Felspulver, der als ein graues Gebiet um die kleinen Eiskristallen dargestelltwird, welche die Produkte der Neukristallbildung der Schur sind

ABB. 4. Der Felsring, der einen Durchmesser von 7 cm hattenachdem er unter dem Eis zu −6°C mit eine Last von 2 kg cm^{-2} und mit einer ‚Stick-slip‘ Bewegung geglitten war. Die dunklen Flecken sind das Eis, das am Felsen klebt

ABB. 5. Die Sandsteine nachdem sie unter dem Eis hin- und hergeglitten sind die ‚Stick-slip‘ Bewegung. Das Eis, das klebt, steht im nichtigen Verhältnis zur Grösse der Partikeln der verschiedenen Sandsteine

ABB. 6. Das Schichtenprofil (eine Breite von 15 cm) von vier Kieschichten nach sieben Kreisläufe eines Frieren- und Tauenprozesses, die mit vier Schichten von sehr feinem Sand abwechseln. Man merkt am meisten eine relative Aufwärtsbewegung, eine Fusion von den zwei obrigen Schichten und das Zerschmettern von den rauheren Partikeln

Deglaciation of the Scoresby Sund fjord region, north-east Greenland

SVEND FUNDER

Institute of Historical Geology and Palaeontology, University of Copenhagen

Revised MS received 1 November 1971

ABSTRACT. The Scoresby Sund region contains some of the world's deepest fjords. The distribution of ice in the region during the Weichselian maximum is not yet known, but in Late Weichselian time the coastal parts were ice-free.

For the Post glacial period, four phases of deglaciation have been distinguished: (1) The Milne Land Stage (?Younger Dryas and Pre-Boreal) comprises some distinct advances of the fjord glaciers; the advances were followed by a period of recession. (2) In the Rødefjord Stage (7500–6700 BP), the fjord glaciers receded intermittently; probably the recession was interrupted by a short period of advance at the end of the period. (3) The Post-glacial warm period (c. 7000–5000 BP) was a time when the glaciers retreated behind their present limits. A warm climate is indicated by the presence of the now-extinct bivalves *Mytilus edulis* and *Chlamys islandica* in raised marine deposits. (4) Historical time is marked by a re-advance.

The deglaciation of the region was greatly influenced by relief. This is reflected in the location of the glacier fronts during the major stages and the differences in the behaviour of neighbouring fjord glaciers. The fjord morphology probably also explains some of the differences between the history of deglaciation in this region and that in West Greenland.

THIS paper deals with some investigations of the Quaternary geology of the Scoresby Sund region, north-east Greenland, carried out by the author in the summers of 1969 and 1970 for the Geological Survey of Greenland (GGU).

RELIEF AND CLIMATE

The Scoresby Sund region is defined as the land area of north-east Greenland draining into the wide 'sound' and its ramifications (Fig. 1). The innermost part of the region is dominated by high mountain plateaux with gently undulating surfaces 1500–2000 m above sea level (Fig. 2). The plateaux are composed of crystalline rocks and carry independent ice caps. H. W. Ahlmann (1941) suggested that the plateaux represent a late Tertiary peneplain. Mountain areas with alpine relief occur in the Stauning Alper in the northern part of the region and near the coast in Liverpool Land, both areas of crystalline rocks. Extensive areas of hilly lowland are found in Jameson Land and parts of southern Milne Land where the bedrock comprises readily weathered Mesozoic sediments.

The present climate in Scoresby Sund is high arctic with a mean annual temperature of 7°C at Scoresbysund settlement (L. Lysgård, 1969). The climate becomes distinctly more continental towards the west, the innermost part of the region having warmer and drier summers.

The Fjords

The Scoresby Sund region is noted for its spectacular fjords, some of the world's longest and deepest. The great depths were first demonstrated by G. Thorson (1934), who made

33

echo soundings of many of the fjords; his data have largely been confirmed by soundings made during the GGU expeditions. The deepest and narrowest fjords are those that trend generally east to west. These fjords may be up to 1400 m deep and have the greater part of their floors at depths of about 1000 m. Taking the mountain sides into consideration, the total relief of the troughs is of the order of 2500–3500 m; their width is about 5 km. These fjords are all cut into crystalline bedrock. Contrary to the observations by Thorson, the soundings made during the GGU expeditions have so far failed to show the presence of any marked thresholds at the mouths of the deep fjords. The fjords with a general north-south trend are generally shallower and wider; their depths do not exceed 600 m. To this group belong the Scoresby Sund-Hall Bredning basin and Rødefjord in the innermost part of the region.

It should be mentioned that the fjords with a general north-south trend are all located along major fault zones with vertical displacements; along the east-west trending fjords, however, no pronounced movement or fracturing seems to have taken place.

Calving glaciers which are outlets from the Inland Ice are found at the heads of many of the fjords. These glaciers and their former extensions are the main subject of the account which follows.

NATURE OF THE EVIDENCE

One object of the investigations was to establish a chronology for the deglaciation of the region. This has been assessed by dating former ice-margin deposits with respect to their intersection by raised marine shorelines, and dating bivalve shells associated with the latter by the radiocarbon method.

During the older phases of the deglaciation moraine ridges were formed at the ice margin; these extend unbroken for 10–15 km along the fjords at altitudes up to 500–600 m. Frontal deposits are rare since the glaciers mainly terminated in sea water. However, in some valleys in Milne Land, terminal moraines of submarine origin have been observed. These occur as rounded banks up to 50 m high composed of sorted sand and gravel with boulders scattered on their top. Where the transition from sub-aerial to submerged moraine is preserved a reliable sea-level date for the formation of the moraine can be obtained (S. Funder, 1970).

During subsequent phases the dominant ice-margin deposits were kame terraces. The terraces can be traced for tens of kilometres along the fjords in the innermost part of the region at levels up to 300 m (Funder, 1971a).

The shorelines upon which the relative chronology is founded are horizontal boulder lines or cliffs in loose deposits that can be followed unbroken for up to 2 km. Often former streams built their deltas to corresponding altitudes and deposits of grey silt may be found sloping down from the beach deposits; the silt deposits may contain bivalve shells *in situ*. Since no shells occur in the beach deposits proper, the shorelines are dated by reference to these shells. During the collection of the samples for dating, a gradation from the silt into a coarse beach deposit 2–4 m higher was observed, the altitude of which was taken to be the sea level at the time that the bivalves lived.

Table I shows all available radiocarbon dates from the region. The location of dated samples, as well as the extension of the major glaciers during different phases of deglaciation is shown in Figure 1. Altitudes of the shore features have been measured using a Paulin barometer, theodolite or hand-level.

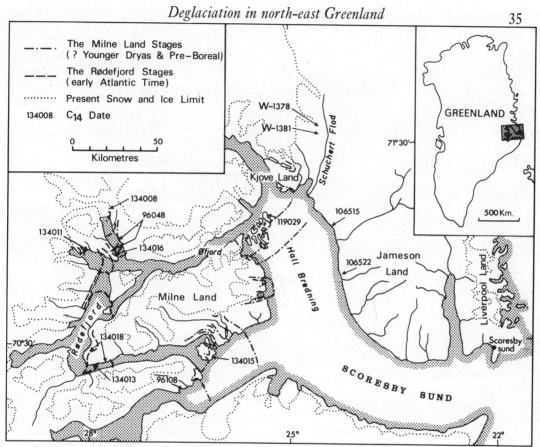

FIGURE I. Deglaciation stages and the location of C¹⁴ dated samples in the Scoresby Sund region. (Information for Kjove Land from Sugden and John, 1965)

THE HISTORY OF DEGLACIATION

Maximum extent of ice cover

Whether the Scoresby Sund region was completely covered with ice during the last ice age is not known. From geomorphological evidence O. Nordenskiöld (1907, p. 277) suggested that the southern part of Jameson Land and the mountain peaks of Liverpool Land were ice-free, while a great ice stream occupied the Scoresby Sund basin and calved into the Greenland Sea. It was subsequently demonstrated, however, that a moraine cover with many far-travelled boulders occurs in large areas of southern Jameson Land (A. Rosen-krantz, 1929, p. 150; H. Aldinger, 1935, p. 27). It therefore seems justified to conclude that this area and probably the whole of the region was once ice-covered; the age of this cover is, however, unknown.

The Milne Land stage (?Younger Dryas and Pre-Boreal)

Moraine ridges and fluvioglacial deposits in eastern and southern Milne Land give evidence of some early stages in the deglaciation of the region (Fig. 1). The moraines were deposited by ice streams which filled the deep east-west trending fjords, and whose fronts lay at the western side of the Scoresby Sund-Hall Bredning basin. Frequently, several parallel

FIGURE 2. View along Øfjord towards the east. The fjord is 1000 m deep and its walls are 1500 m high. It is cut into a high mountain plateau. (Reproduced with permission (A. 533/71) from the Geodetic Institute of Copenhagen)

moraine systems occur together and by dating the individual systems by reference to shorelines, it has been possible to trace an older and a younger phase in all parts of the region; however, it is not known whether the two phases represent one period of oscillatory retreat or two distinct periods of advance.

During the older phase, which is dated by shorelines 110–130 m above sea level in Milne Land, Scoresby Sund and Hall Bredning were bordered to the north and west by cliffs of ice representing the merged fronts of glaciers from all fjords and valleys. In Kjove Land moraine ridges indicating the presence of glaciers in the adjacent fjord and in the Schuchert Flod valley were reported to be older than a shoreline 134 m above sea level (D. E. Sugden and B. S. John, 1965): these ridges could probably be equivalent to this older phase.

The younger phase is dated by shorelines 90–95 m above sea level in Milne Land. During this phase the glaciers had receded and taken up positions at the actual mouths of the deep fjords; lobate glacier fronts extended over the islands of Bjørneøer in the north of the region and Danmark Ø in the south. In Kjove Land the last indication of ice streams in the adjacent fjord and in the Schuchert Flod valley is contemporaneous with a shoreline

101 m above sea level (Sugden and John, 1965); this could be equivalent to the younger phase of the Milne Land stage.

The Milne Land stage has not so far been dated absolutely. An idea of its age can be obtained by extrapolation from dated marine deposits at lower levels. Thus the shoreline datings from areas around Hall Bredning (samples W-1381, 106522 and 119029, Fig. 1 and Table I) agree well with those obtained from Mestersvig and Skeldal 100 km to the north of

TABLE I

Radiocarbon dates from the Scoresby Sund region

Sample no.	Locality	Age, C[14] years BP	Corresponding sea level	Material
W-1378[1]	Schuchert Flod valley	1490± 250	–	turf
W-1381[1]	Schuchert Flod valley	7900± 350	50	bivalve shells
96048[2] K-1459	Rypefjord	6800± 130	25?	bivalve shells
96108[2] K-1460	Gåseland	6910± 130	8?	bivalve shells
106515[3] K-1740	Jameson Land	2290± 140	–	lake mud
106522[3] K-1741	Jameson Land	8580± 140	c. 53	lake mud
119029[2] K-1461	Bjørneøer	8640± 140	65	bivalve shells
134008[3] I-5420	Rypefjord	6650± 125	36	bivalve shells
134011[3] I-5421	Harefjord	7140± 130	50	bivalve shells
134013[3] I-5422	Fønfjord	6450± 120	25	bivalve shells
134015[3] I-5423	Danmark Ø	6840± 125	6?	bivalve shells
134016[3] K-1742	Rypefjord	6200± 140	–	lake mud
134018[3] K-1743	West Milne Land	6780± 140	–	lake mud

[1] B. Levin *et al.*, 1965; [2] H. Tauber, 1970; [3] S. Funder, 1971b. Sample numbers without a letter are GGU sample numbers.

Hall Bredning (A. L. Washburn and M. Stuiver, 1962; N. Lasca, 1969), which makes it probable that the areas lie near the same isobase of uplift. For Mestersvig, Washburn and Stuiver (1962) calculated an initial rate of uplift for the period prior to 9000 years BP of 9 m/100 years. For the neighbouring Skeldal, Lasca (1969) gave a rate of 3 m/100 years for the period 8000–7000 BP. If these values can be regarded as maximum and minimum possibilities for the early uplift around Hall Bredning, the true age of the older phase of the Milne Land stage must lie within the range 11 000–9500 BP, while for the younger phase it lies between 9500 and 9000 BP.

The Rødefjord stage (early Atlantic)

Succeeding the Milne Land stage at the mouths of the deep fjords, evidence of younger glaciers in the fjords is found in the innermost part of the region (Fig. 1). Along Rødefjord and the fjords to its north and south, extensive systems of kame terraces occur. The predominance of kame deposits over moraines, the lack of synchroneity of the individual stages from fjord to fjord, and the location of the glacier fronts at places where the fjords branch or where the inclination of their sides changes from steep to gentle, indicate that the deposits were laid down during minor halts, conditioned by morphology, in a period of down-wasting of the glaciers. However, near the present glacier fronts in Rypefjord and Harefjord, to the north of Rødefjord, moraines are found; they seem to reflect a small readvance of these glaciers at the end of the Rødefjord stage.

The oldest deposits referred to this stage are kame terraces found at the mouths of Rypefjord and Harefjord, to the north of Rødefjord, which are contemporaneous with shorelines 60 m above sea level. A radiocarbon date for bivalve shells related to a 50 m shoreline in the area is 7140 BP (sample 134011, Fig. 1 and Table I); extrapolation from this and younger dates from the area indicate an age of about 7500 years BP for the beginning of the Rødefjord stage.

At this time, Øfjord, 80 km long, had been deglaciated; the average rate of recession of the glacier in this fjord during late Pre-Boreal, Boreal and early Atlantic time was about 40 m/year. However, in the neighbouring Fønfjord, to the south of Milne Land, shorelines and radiocarbon dating indicate a much slower deglaciation. Kame terraces found near the head of Fønfjord where it meets Rødefjord are contemporaneous with shorelines 38 m above sea level. The dating of sample 134013 (Fig. 1 and Table I) indicates that the rate of uplift in this area was nearly the same as that at the head of Øfjord, and the date of 6780 BP for the base of organic sediments in a lake immediately inside deposits of the Rødefjord stage (sample 134018, Fig. 1 and Table I) indicates that Fønfjord was deglaciated at least 500 years later than Øfjord, the average rate of recession in the former being only about 20 m/year. The most likely explanation for the difference in the rates of recession of the glaciers in the two fjords lies in their morphology; this will be discussed later (p. 40).

The end of the Rødefjord stage was chosen to mark the time when deglaciation reached approximately its present extent; this took place a short time prior to the formation of shorelines at 35 m above sea level; a shoreline at this altitude can be found immediately in front of many of the existing calving glaciers. In Rypefjord, to the north of Rødefjord, this shoreline is also found along the margin of the present glacier behind its front, indicating that at the time of its formation the glacier was behind its present limits. A radiocarbon date for bivalve shells related to this shoreline gave 6650 BP, providing a minimum date for the end of the Rødefjord stage (sample 134008, Fig. 1 and Table I).

It can be concluded that the Rødefjord stage was a time of down-wasting of the major glaciers in the region, probably with minor readvances at its close. The period lasted from about 7500 to 6700 BP; however, owing to the character of the deposits the boundaries must be considered somewhat arbitrary.

The Post-glacial warm period (late Atlantic)

In north-east Greenland, a Post-glacial period with a climate warmer than the present was first demonstrated by A. G. Nathorst (1901, p. 304), who found shells of the now-extinct bivalve species *Mytilus edulis* in raised marine deposits as far north as 73° N. The present northern limit for *Mytilus* in East Greenland waters is at 66° N (W. K. Ockelmann, 1958). Subfossil *Mytilus* has later been found at several localities to the north of Scoresby Sund (A. Noe-Nygård, 1932), and during the present investigations it was also found in

Scoresby Sund at several localities up to 35 m above sea level. A second species that probably immigrated to the Scoresby Sund region and probably became extinct together with *Mytilus* is *Chlamys islandica*. This species, it is true, has been found living in Kong Oscars Fjord to the north of Scoresby Sund (Thorson, 1933), but since it occurs subfossil only together with *Mytilus*, and since its northern limit south of Kong Oscars Fjord is at 61° N, 1000 km to the south (Ockelmann, 1958), it should probably be regarded as a relict form in its present habitats in Kong Oscars Fjord.

It has been mentioned (p. 38) that in Rypefjord, to the north of Rødefjord, a shoreline and a raised marine deposit are found behind the margin of the present glacier; the deposit contains *Mytilus* shells and extends to 1500 m behind the present front of the glacier (site for sample 134008, Fig. 1). The glacier therefore must have advanced over deposits of the warm period from a position at least 1500 m behind its present front, and probably more, since it seems unlikely that *Mytilus* would have lived in the muddy water at the front of a glacier.

An observation made by Hartz (in E. Bay, 1896) indicate that the same may be true for other glaciers in the region. Hartz found shell fragments of *Chlamys islandica* on the surface of Rolige Bræ, a calving glacier in the southern part of Rødefjord. Probably this glacier has advanced over marine deposits from the warm period, and shearing in the glacier brought them to its upper surface.

Bivalve faunas containing *Mytilus* have been dated from three localities in the Scoresby Sund region; the dates are 6840, 6650 and 6450 BP (samples 134015, 134008 and 134013, Fig. 1 and Table I).

In the Mestersvig and Skeldal areas, *Mytilus* occurs in the identification lists for four dated samples; the ages for the samples lie in the range 6960–5680 BP (Washburn and Stuiver, 1962; Lasca, 1969). If the correction of ± 550 years used by Washburn and Stuiver is also applied to the ages from Skeldal, then the total range would be 6840–5130 BP.

The good agreement between the seven datings indicates that *Mytilus* probably lived in the region only during this period. If taken as an indicator for the warm period, i.e. a period with a climate warmer than the present, then the minimum duration of this period was from 7000 to about 5000 BP.

Younger periods

The end of the Post-glacial warm period is indicated in the author's pollen diagrams from Jameson Land by a distinct rise in the pollen percentages of some high arctic species. Whether this took place at the same time as the disappearance of *Mytilus* from the region is not yet established. The subsequent deterioration of the climate reached a maximum in historical time; fresh moraines without vegetation occur some distance in front of the present glacier snouts. The moraines have not been precisely dated, but their age probably docs not exceed a few hundred years.

A very recent advance of some glaciers in the Stauning Alper has been described by O. Olesen and N. Reeh (1969); some of these glaciers are known to be surging.

SUMMARY OF THE HISTORY OF DEGLACIATION

Deglaciation of the Scoresby Sund region was already far advanced by Late Weichselian time. The coastal areas, the Scoresby Sund-Hall Bredning basin and a great part of the adjacent Jameson Land were probably ice-free prior to 11000 BP. In Pre-Boreal and possibly

Younger Dryas time some major advances—the Milne Land stage—took place. Glaciers in all fjords and valleys advanced to the northern and western side of Scoresby Sund where they merged to form calving walls of ice. Before the end of Pre-Boreal time, ice recession accelerated and the deep and narrow fjords were opened to the sea. The average rate of recession of the major fjord glaciers in this period was 40–20 m/year. During the further recession in the innermost part of the region, where the fjords are shallower and their sides less steep, several halts conditioned by morphology took place (the Rødefjord stage); after a minor readvance, the present state of deglaciation was achieved about 6700 years ago, and in the following period the glaciers receded behind their present limits; how far is not known. During the same period the 'southern' bivalves *Mytilus edulis* and *Chlamys islandica* migrated into the region, to become extinct again probably about 5000 years ago. The subsequent climatic deterioration reached its maximum in historical time.

DISCUSSION

Some effects of fjord morphology upon the behaviour of calving glaciers have been described by J. Mercer (1961) who showed that a glacier in an 'ideal' fjord, i.e. a linear fjord of constant width, will react to a vertical change in the firn limit by advancing or retreating until the terminus reaches a point where the fjord ceases to be 'ideal', as where the fjord widens or bifurcates, or where the inclination of its sides changes abruptly; only at such places can equilibrium in the glacier be re-established. Therefore, according to local relief, the calving glaciers in a fjord region may move out of phase with each other.

Mercer's observations have a strong bearing on the course of deglaciation in Scoresby Sund. Figure 1 demonstrates that the fronts of the fjord glaciers during all major halt or advance stages were located at places where the fjords branch, or at their mouths, which indicates the importance of the morphology in determining the former extensions of these glaciers. Further, it has been demonstrated (p. 38) that in the neighbouring fjords Øfjord and Fønfjord, to the north and south of Milne Land, the rates of recession of the glaciers differed markedly; the average figures are, respectively, 40 and 20 m/year for late Pre-Boreal, Boreal and early Atlantic times. Øfjord (Fig. 2) is very close to being an 'ideal' fjord: along the 80 km of its course the fjord is of uniform width, its walls are steep and no major tributary valley enters the fjord. Fønfjord, by contrast, is of a more irregular outline, its walls varying from steep to gentle and several major valleys joining the fjord. It seems probable that the delay of more than 500 years in the deglaciation of Fønfjord is a consequence of its morphology.

In a correlation of the stages of deglaciation in Greenland, A. Weidick (in press) pointed out that deglaciation apparently began earlier in the extensive fjord zone of northeast Greenland than was the case in west Greenland, and that, in west Greenland, fjord regions such as the Godthåb Fjord and Disko Bugt areas were deglaciated earlier than adjacent areas with few fjords such as the large area draining into Søndre Strømfjord. In explanation he suggested that the fjord morphology allowed rapid drainage from the Inland Ice margin during the melting phases at the end of the Wisconsinan/Weichselian.

A further marked difference pointed out by Weidick concerning the progress of deglaciation on the two sides of Greenland is that the 'Fjord stages' known from west Greenland have not been traced in east Greenland. The Fjord stages (Weidick, 1968; in press) consist of some marked advance or halt stages that affected large areas along the margin of the Inland Ice in west Greenland in Boreal time. Since an oscillation of this age has not been recorded either for fjord glaciers or for major valley glaciers in north-east Greenland, this difference is probably not conditioned by morphology. However, it appears that within the general framework of climatic change proposed here, fjord morphology was a most decisive factor during the deglaciation of the Scoresby Sund region.

ACKNOWLEDGEMENTS

Thanks are due to the director of the Geological Survey of Greenland for permission to publish these results, to mag. scient. Niels Henriksen, leader of the Scoresby Sund expeditions, for much valuable advice and to Dr A. K. Higgins for improving the English.

REFERENCES

AHLMANN, H. W. (1941) 'The main morphological features of north-east Greenland', *Geogr. Annlr* 23, 148–83

ALDINGER, H. (1935) 'Geologische Beobachtungen im oberen Jura des Scoresbysundes', *Meddr Grønland* 99, 1, 128 pp.

BAY, E. (1896) 'Geologi', *Meddr Grønland* 19, 145–89

FUNDER, S. (1970) 'Notes on the glacial geology of eastern Milne Land', *Rapp. Grønl. geol. Unders.* 30, 37–42

FUNDER, S. (1971a) 'Observations on the Quaternary geology of the Rødefjord region, Scoresby Sund, East Greenland', *Rapp. Grønl. geol. Unders.* 37, 51–5

FUNDER, S. (1971b) 'C14 dates from the Scoresby Sund region, 1971', *Rapp. Grønl. geol. Unders.* 37, 57–9

LASCA, N. (1969) 'The surficial geology of Skeldal, Mesters Vig, north-east Greenland', *Meddr Grønland* 176, 3, 59 pp.

LEVIN, B., P. C. IVES, C. L. OMAN and M. RUBIN (1965) 'U.S. Geological Survey radiocarbon dates VIII', *Radiocarbon* 7, 396

LYSGÅRD, L. (1969) 'Foreløbig oversigt over Grønlands klima', *Meddr dansk met. Inst.* 21, 35 pp.

MERCER, J. (1961) 'The response of fjord glaciers to changes in the firn limit', *J. Glaciol.* 3, 850–8

NATHORST, A. G. (1901) 'Bidrag til Nordöstra Grönlands geologi', *Geol. För. Stockh. Förh.* 23, 275–306

NOE-NYGÅRD, A. (1932) 'Remarks on *Mytilus edulis L.* in raised beaches in east Greenland', *Meddr Grønland* 95, 2, 24 pp.

NORDENSKIÖLD, O. (1907) 'On the geology and physical geography of east Greenland', *Meddr Grønland* 28, 151–285

OCKELMANN, W. K. (1958) 'Zoology of east Greenland, marine lamellibranchiata', *Meddr Grønland* 122, 257 pp.

OLESEN, O. and N. REEH (1969) 'Preliminary report on glacier observations in Nordvestfjord, east Greenland', *Rapp. Grønl. geol. Unders.* 21, 41–53

ROSENKRANTZ, A. (1929) 'Preliminary account of the geology of the Scoresby Sund district', *Meddr Grønland* 173, 2, 133–54

SUGDEN, D. E. and B. S. JOHN (1965) 'The raised marine features of Kjove Land, east Greenland', *Geogrl J.* 131, 235–47

TAUBER, H. (1970) 'C14 dates on Post-glacial marine shells' in 'Report on the 1969 geological expedition to Scoresby Sund, east Greenland', *Rapp. Grønl. geol. Unders.* 30, 43–4

THORSON, G. (1933) 'Investigations on shallow water animal communities in the Franz Joseph Fjord (east Greenland)', *Meddr Grønland* 100, 2, 69 pp.

THORSON, G. (1934) 'Contributions to the animal ecology of the Scoresby Sound Fjord complex (east Greenland)', *Meddr Grønland* 100, 3, 69 pp.

WASHBURN, A. L. and M. STUIVER (1962) 'Radiocarbon-dated post-glacial delevelling in north-east Greenland and its implications', *Arctic* 15, 66–74

WEIDICK, A. (1968) 'Observations on some Holocene glacier fluctuations in West Greenland', *Meddr Grønland* 165, 3, 202 pp.

WEIDICK, A. (in press) 'Holocene shorelines and glacial stages in Greenland—an attempt at correlation', *Rapp. Grønl. geol. Unders.* 41

RÉSUMÉ. *L'histoire de la déglaciation dans la région de Scoresby Sund, Groenland de l'est.* La région de Scoresby Sund est une zone de fjords étendue contenant quelques-uns des fjords les plus profonds de la terre. L'extension de la glace pendant la glaciation Weichselienne n'est pas encore connue, cependant on sait que la partie côtière de la région était libre de la glace vers la fin de cette période.

En ce qui concerne la période Post-glaciaire, les phases suivantes de déglaciation ont pu être distinguées: (1) Le stade de Milne Land (? Dryas récent et Pré-Boréal); quelques distinctes avances des glaciers des fjords suivies par un rapide recul. (2) Le stade du Rødefjord (7500–6700 ans); ceci comprends apparemment des stades d'arrêt controlés essentiellement par la morphologie pendant le retrait des glaciers des fjords; éventuellement il y en quelques réavancements à la fin de la période. (3) La période chaude Post-glaciaire (au moins 7000–5000 ans); pendant cette période les glaciers étaient en arrière de leurs limites actuelles. Un climat chaud dans la région est indiqué par le présence des bivalves maintenant extincts *Mytilus edulis* et *Chlamys islandica* dans les dépots marins surélevés. (4) Pendant la période suivante, les glaciers des fjords avancèrent encore une fois jusqu'à un maximum durant les temps historiques.

L'histoire de la déglaciation de Scoresby Sund était influencée par la morphologie des fjords; ceci pourrait expliquer les différences entre les évènements quaternaires des côtes ouest et est du Groenland.

FIG. 1. Stades de déglaciation et datations C14 dans la région de Scoresby Sund

FIG. 2. Vue le long du Øfjord vers l'est. Le fjord est entouré par des parvis de 1500 m de hautes surmontées par un haut-plateau

---

ZUSAMMENFASSUNG. *Enteisungsgeschichte in der Scoresby Sund Region, Ost Grönland.* Die Scoresby Sund Region ist ein weitläufiges Fjordsystem, das einige der tiefsten Fjorde der Welt enthält. Die maximale Ausbreitung des Eises während der Weichsel Eiszeit ist noch nicht bekannt, die Küstenregionen der Gegend waren jedoch im Spätglazial eisfrei.

Folgende vier Phasen der Enteisung wurden für die Post-glaziale Periode unterschieden: (1) Die Milne Land Stadium (? Jüngere Dryas und Prä-Boreal) einschliesst einige deutliche Vorstösse der Fjordgletscher, gefolgt von einem schnellen Rükzug. (2) Die Rødefjord Stadium (7500–6700 Jahre v.H.) scheint hauptsächlich morphologisch kontrollierte Halt-Stadien während des Rückzuges der Fjordgletscher zu sein; möglicherweise traten einige kleinere Vorstösse am Ende dieser Periode auf. (3) Die Post-glaziale Wärmezeit (mindestens 7000–5000 Jahre v.H.), während dieser Periode waren die Gletscher hinter ihren jetzigen Grenzen. Ein warmes Klima ist durch die Präzens der jetzt hier ausgestorbenen Muscheln *Mytilus edulis* und *Chlamys islandica* in gehobenen Meeresablagerungen angezeigt. (4) In der folgenden Zeit stiessen die Fjordgletscher nochmals vor, bis sie in historischer Zeit ein Maximum erreichten.

Die Enteisungsgeschichte des Scoresby Sundes ist durch die Fjordmorphologie stark beeinflusst. Dies mag auch als Erklärung für einige Unterschiede im Verlauf der Enteisung in West- und Ostgrönland dienen.

ABB. 1. Enteisungsstadien und C14-Datierungen in der Scoresby Sund Region

ABB. 2. Sicht entlang dem Øfjord gegen Osten. Der Fjord ist 1000 m tief und durch 1500 m hohe Wände begrenzt, auf denen sich ein Hochgebirgsplateau befindet

---

The contribution of radio echo sounding to the investigation of Cenozoic tectonics and glaciation in Antarctica

DAVID J. DREWRY

Scott Polar Research Institute, Cambridge

MS received 13 October 1971

ABSTRACT. Continuous radio echo profiling of the ice/bedrock interface in Antarctica has been undertaken from an airborne platform by the Scott Polar Research Institute and the National Science Foundation of the United States during the austral summers of 1967–68 and 1969–70.

The method utilizes a pulse-modulated 35MHz radio sounding system having a resolution of 10 m in ice. The system is many times more rapid and is probably more accurate in deep polar ice than determinations of thickness by seismic shooting. The profiling technique has provided new evidence of the recent diastrophic and glacial history of Antarctica.

Radio echo soundings along the inland side of the Transantarctic Mountains from Victoria Land southwards to the Queen Maud Mountains confirm recent surface geological investigations and indicate a complex pattern of differentially tilted fault-blocks. Some of these blocks, possibly comprising Beacon Supergroup rocks, extend up to 600 km from the Ross Sea coast. Variations in the magnitude of tilting and secondary longitudinal and transverse faulting within the blocks combined with varying amounts of erosion have produced a complicated transitional zone between the epicratonic mountain belt and the lowland shield of the East Antarctic craton.

On the inland side of the mountains, valleys of probable glacial origin, but now submerged beneath the ice sheet, indicate that glaciers once descended both flanks of the Transantarctic Mountains. It is thought that these troughs, together with those at present penetrating the mountains, were eroded by local mountain glaciers during the mid-Cenozoic, following the initial uplift of the Transantarctic Mountain belt. With the onset of the full-scale continental glaciation the lower, inland sector became submerged by the progressive accumulation of ice in the interior. The present ice-flow pattern of the East Antarctic ice sheet is consequently discordant with much of the sub-glacial relief. This buried relief may have been little modified by recent cold-based ice.

CONTINUOUS radio echo sounding from aircraft has been developed in recent years by a team at the Scott Polar Research Institute (S.P.R.I.), Cambridge. It has made possible the rapid profiling of the ice/bedrock interface over vast areas of Antarctica (J. T. Bailey, S. Evans and G. de Q. Robin, 1964; Evans and Robin, 1966; Evans, 1967; Robin, C. W. M. Swithinbank and B. M. E. Smith, 1970; Robin *et al.*, 1970). The data used in this report were obtained during the austral summers of 1967–68 and 1969–70 by teams from the S.P.R.I. with the logistic support of the National Science Foundation (N.S.F.) of the United States. The author participated in the field during the 1969–70 Antarctic season but several other people have contributed substantially to obtaining the echo records used here, in terms of the development and operation of equipment.

Geomorphological interpretation of the data obtained from soundings inland of the Transantarctic Mountains between the Byrd Glacier and the Queen Maud Mountains reveal enough detail to evaluate and partially to reconstruct the sub-ice structure of this part of Antarctica. In addition, the detailed morphology of the sub-ice reflections provides details of the incipient stage of Antarctic glacierization.

THE RADIO ECHO SOUNDING SYSTEM

Continuous radio echo profiling is not only more rapid but may surpass in accuracy conventional geophysical techniques, such as gravimetric and seismic methods (the former usually being tied to the latter), in determining depth to bedrock through cold, polar ice. This is primarily a result of the fact that both radio echo sounding and seismic shooting are sensitive, in contrasting ways, to temperature changes in the sounding medium. Robin (1958) and A. P. Crary (1963) have shown that, in Polar seismic exploration, cold ice ($> -30°$ to $-40°$C) can result in substantial noise from incoherent surface waves. The echoes then become extremely difficult to identify and to interpret. Many seismic results obtained on early traverses in East Antarctica suffered in this way and many of the seismograms showed no unambiguous reflections (C. R. Bentley, 1964).

The performance of radio echo sounding is also strongly dependent upon but inversely related to temperature (Robin, 1972). This is caused by increasing dielectric absorption per metre path length as ice temperatures approach the melting point. Laboratory experiments on ice obtained from a Greenland core show that losses at $-1°$C are approximately 0.05dB m^{-1} and 0.001dB m^{-1} at $-60°$C (Robin, Evans and Bailey, 1969). The performance of the system in respect of power losses by absorption is considerably better in the very cold ice of the East Antarctic plateau, just where seismic shooting encounters the greatest difficulties. Seismic refraction profiles, however, are still invaluable in providing information on velocity zonation in bedrock sections.

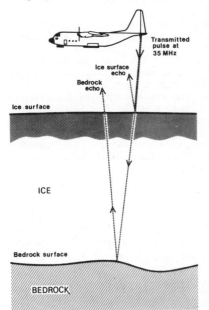

FIGURE 1. The Scott Polar Research Institute 35MHz, airborne radio echo sounding system. Electromagnetic energy is radiated downwards from antennae mounted in the tail of a U.S. Navy Hercules C-130B aircraft. The polar diagram indicates a beamwidth of 25° in the fore-and-aft direction and 10° in the transverse direction. Echo returns are obtained from ice/air, ice/rock, and ice/water interfaces

Description of the system

Similar to marine echo sounding techniques, the S.P.R.I. radio echo sounding system directs a pulse of transmitted energy downwards to the ice surface from an oversnow vehicle or an airborne platform (Fig. 1). The delay of the returning echo indicates the range to the reflecting surface (Fig. 2).

Radio frequencies between 30MHz and 500 MHz have been used since dielectric absorption increases very rapidly beyond 500MHz reaching 10dB per 100 m path length at 1000 MHz. The S.P.R.I. Mark IV system operates at 35 MHz and ranges in ice to an accuracy of 10 m can be resolved. Pulses returning from the ice surface and sub-ice reflectors are displayed on an intensity modulated cathode-ray tube (Fig. 2) and integrated on a single record by a continuously moving photographic film. Calibration marks are inserted electronically every minute and identified by an independent character string (see Figure 7 as an example).

Several factors have been shown to be important in the performance of radio echo sounding (Robin, Evans and Bailey, 1969): system performance (ratio of trans-

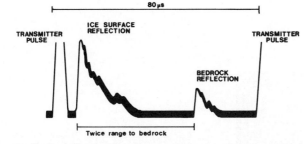

FIGURE 2. An A-scope display indicates salient features during ice-thickness measurements. The S.P.R.I. Mark IV System has a delay of 80 μs between transmitter pulse triggers. Returning reflections from ice or sub-ice surfaces are shown. The irregular echo tail results from scattering of energy over rough reflecting surfaces

mitted power to the minimum detectable power at the receiver); dielectric absorption of radio waves in ice; geometrical factors (size and configuration of the antennae, spreading of radio waves from a point and refraction effects at the snow surface) and finally reflection losses (ratio of incident energy to the reflected energy from the sub-ice surface). Calculations of echo strength have been made for several locations in Greenland and Antarctica (Robin, Swithinbank and Smith, 1970). They show that good echoes should be obtained over most of the Antarctic ice sheet excluding parts of Marie Byrd Land and the deepest areas of East Antarctica.

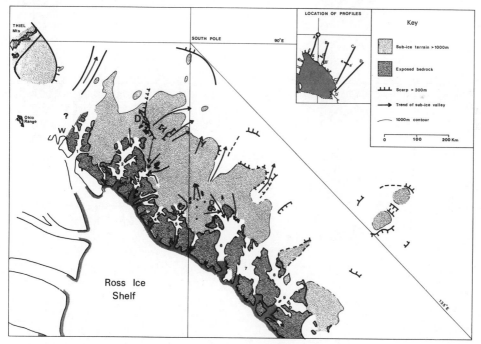

FIGURE 3. Sub-glacial structural elements between the southern Transantarctic Mountains and the South Polar plateau. Sub-glacial terrain above 1000m is indicated and the trend of sub-glacial valleys is shown. The numbered outlet glaciers are: 1. Reedy Glacier; 2. Robert Scott Glacier; 3. Amundsen Glacier; 4. Liv Glacier; 5. Shackleton Glacier; 6. Beardmore Glacier; 7. Lennox-King Glacier; 8. Marsh Glacier; 9. Nimrod Glacier. W = Wisconsin Range; D = D'Angelo Bluff; O = Otway Massif

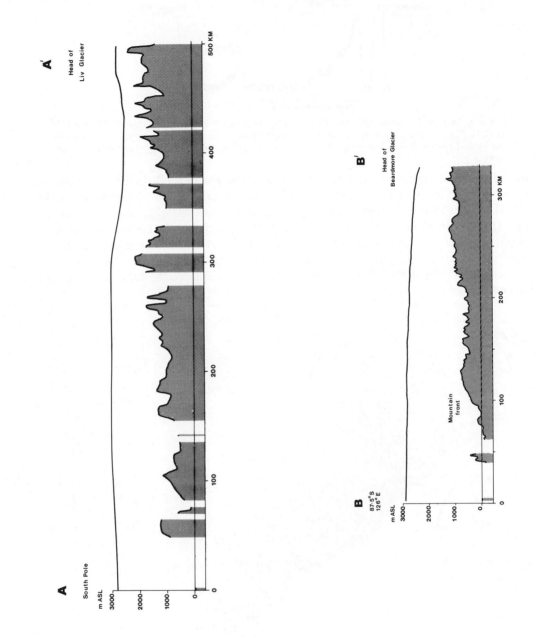

FIGURE 4. Real space profiles obtained by continuous radio echo sounding between the East Antarctic plateau and the Transantarctic Mountains. The location of the profiles is shown in Figure 3.

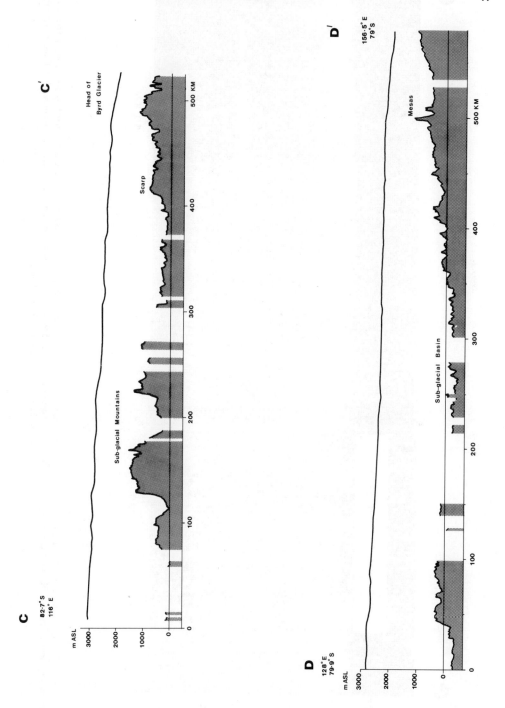

FIGURE 5. Real space profiles obtained by continuous radio echo sounding between the East Antarctic plateau and the Transantarctic Mountains. The location of the profiles is shown in Figure 3.

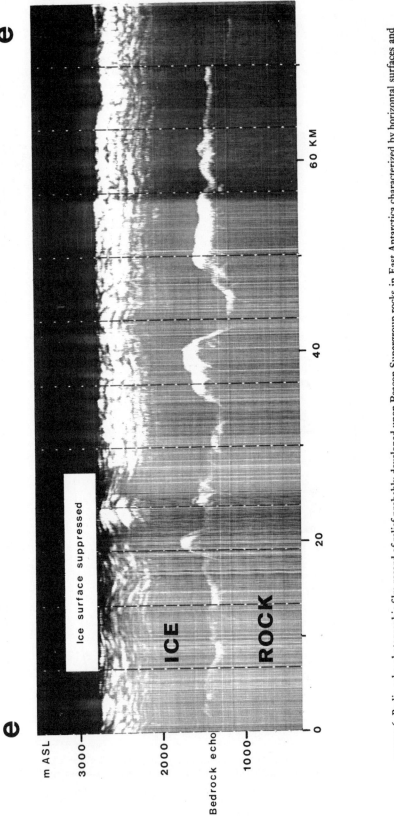

FIGURE 6. Radio echo photographic film record of relief probably developed upon Beacon Supergroup rocks in East Antarctica characterized by horizontal surfaces and steep escarpments. The film, obtained during 1967–68 operations, is printed with the horizontal scale compressed and the vertical scale expanded. Echoes returning from the ice/bedrock interface are shown

Operations and data reduction

During the 1967–68 and 1969–70 seasons, a joint S.P.R.I./N.S.F. programme of radio echo sounding concentrated several flights through and adjacent to the Transantarctic Mountains into East Antarctica, from southern Victoria Land to the Queen Maud Mountains (Fig. 3) (M. Forbes, 1968; Evans and Smith, 1970; Robin *et al.*, 1970).

There are considerable problems in analysing the results obtained during these flights. Flight tracks of the aircraft are obtained from the reduction of SFIM flight-recorder traces. A graphic output of aircraft track is produced by computer. The track is updated with independent fixes derived from the photogrammetrical resection of high oblique and vertical aerial photographs, obtained with a Trimetrogon System. This method of fixing aircraft position can only be undertaken when within sight of known topographical features but is accurate to within ±2 km.

Radio echo film records of ice thicknesses are reduced by converting the echo time delay to ice depths. A propagation velocity for radio waves in ice of 169 m μs^{-1} is used in this conversion. The determination of this value from laboratory and field experiments is discussed in Robin, Evans and Bailey (1969) and J. W. Clough and C. R. Bentley (1970). The error introduced by variations of temperature between 0°C and −60°C and by the effects of crystal orientation in an ice sheet is of the order of ±3 m μs^{-1}.

Ice-surface elevations are obtained from single airborne altimetry corrected to ground control altitudes determined by multiple altimetry during oversnow traverses, and photogrammetry (with primary ground survey) in the vicinity of the Transantarctic Mountains (as depicted on the U.S.G.S. 1:250 000 Reconnaissance Series Maps).

SUB-GLACIAL RELIEF AND STRUCTURE

The results of the lengthy data-handling process have been used to build up a picture of the sub-glacial relief between the Transantarctic Mountains and the South Pole from the Byrd Glacier to the Thiel Mountains, in the sector between longitude 145° E and 90° W (Fig. 3). Preliminary contour maps of sub-glacial relief have been produced at scales of 1:2 200 000 and 1:500 000, with contour intervals of 500 m and 300 m respectively (Drewry, 1972). These maps and details of radio echo profiles have been used to interpret sub-glacial bedrock structure and geomorphology.

The sub-glacial relief is complex but two major relief zones have been distinguished. Ice, 500–2000 m in thickness, covers an irregular mountainous area whose relief amplitude is of the order of 300 m to 1500 m (at 500 m to 2500 m above sea level) and which extends from a series of intra-montane embayments in the southern Transantarctic Mountains to about 145° E. Beyond this the other major zone is encountered—the undulating interior lowland of East Antarctica where the amplitude of the relief is reduced to about 200–400 m, with wavelengths between 4 and 6 km and at heights of 0–1000 m. High-standing ridges and outliers penetrate farther inland between 169° E and 140° W, and a discrete sub-glacial mountain range, divorced from the main belt of the Transantarctic Mountains, is located in the region around 88° S, 120° E (Fig. 3).

The transition between these two provinces appears to be very variable. In places there is a gradual transition from the mountains to the lowland with a gradual reduction of amplitude and absolute relief inland (Fig. 5, D–D'). Elsewhere, escarpments (300–1000 m in height) mark the inland termination of the mountain zone (Fig. 4, B–B'). Probable up-faulted outliers, beyond the transitional zone, exist as mountain blocks (Fig. 5, C–C').

These variations are interpreted as a result of the differential inland tilting of tectonic fault-blocks which extend along the plateauward side of the Transantarctic Mountains. Such interpretations of the profiling data are considerably assisted by the remarkably well-developed structural control of relief in the mountains, which has resulted, where it has been determined in the exposed sectors, from a combination of Mid- to Late Tertiary block faulting and extensive intrusion by dolerite bodies into a thick (3000 m) sub-horizontal sedimentary sequence—the Beacon Supergroup (P. J. Barrett et al., 1972). The exposed relief is, in consequence, dominated by structural benches, mesas, tablelands and single and multiple escarpments (V. R. McGregor, 1963).

It is thought that the rocks of the Beacon Supergroup succession extend well out into East Antarctica, up to 600 km from the Ross Sea coast at latitude 82°30′ S. Here radio echo evidence indicates that there are scarp-terminated, near-horizontal surfaces developed in bedrock with little surface relief and cut by steep-walled valleys (Fig. 6), indicative of Beacon rocks. Interpretation of the meagre magnetic evidence available also suggests that a sedimentary section exists in this area, although it is probably only thin (E. S. Robinson, 1962). In this case the Beacon succession has probably been preserved by vertical uplift of the strata in the form of outlying mountain ranges shown in Figures 3 and 5 (C-C′).

In other areas the Beacon rocks extend into East Antarctica where the fault-blocks of the Transantarctic Mountains are gently tilted inland and hence outcrop as large ridges. This is the case between the Queen Maud Mountains and the South Pole. A crustal block is tilted a few degrees to the south-west and extends from the head of the Robert Scott Glacier to about 88°30′ S. It also dies out laterally beyond the 180° meridian. The fault-block has a steep scarp face (1000–1500 m high) on its northern and eastern margins (shown in Figure 3 adjacent to D). D'Angelo Bluff appears as a small exposed section of the northern block-edge. Parts of the upland reach over 2000 m in height, but in places the surface comprises a regular tableland at 1800 m with surface relief less than 100 m in amplitude (Drewry, 1972; Fig. 4), and probably reflects structural control by the Beacon Supergroup. Near the South Pole, seismic refraction profiles indicate an upper layer in which the velocities for the compressional wave (3·5 to 4·5 km s^{-1}) were interpreted as passing through sedimentary strata or frozen till (Robinson, 1964; Clough and Bentley, 1972). In view of the attitude and morphology of the bedrock in this region (Fig. 4, AA′), a sedimentary interpretation is preferred. The absence of steep magnetic anomalies further suggests the presence of a thin sedimentary rock layer (Robinson, 1962).

In general terms it appears that a simple tectonic model for the inland flank of the Transantarctic Mountains, such as the faulted horst of T. W. E. David and R. E. Priestley (1914), G. T. Taylor (1930), R. W. Fairbridge (1952) and R. L. Nichols (1966, 1969), can no longer be admitted. The exposed and the sub-glacial relief indicate that there has been a main axis of uplift along a coastal (eastern) fault producing a series of major crustal blocks which are tilted inland (McGregor and A. F. Wade, 1969). In some cases the magnitude of vertical uplift and degree of tilting leads to a gradual reduction of absolute relief towards the interior. In others, where the strata may have been less severely tilted, more uplifted or suffered less denudation, steep escarpments characterize the transitional zone.

GLACIAL INVESTIGATIONS

Investigation in areas adjacent to the present-day exposed Transantarctic Mountains has helped to elucidate the history and mode of glacial activity in the mountain zone.

Radio echo records have indicated the presence of sub-glacial valleys in this area. They appear to be of two types. There are those which are tributary to existing outlet glaciers (for example, Reedy, Robert Scott, Amundsen, Liv, Shackleton and Beardmore Glaciers) and which extend the bedrock networks of these glaciers (Fig. 3). The second type of valley consists of those channels which are entirely sub-glacial, independent of the outlet systems, and which drain in an inland direction (Fig. 3). These valleys appear to be distributed on the plateauward side of several upland blocks beneath the ice sheet. Figure 7 shows a radio echo photograph across three of these valley features, in this case flanked by nunataks. The origin of these valleys is thought to be glacial for the following reasons:

(1) The shape of the valley in cross-section is similar to the morphometry of known glaciated valleys. The shape of the sides of the left-hand valley in Figure 7 is approximated by the following equations:

$$y = 6 \cdot 988 \times 10^{-7} \, x^{2 \cdot 273} \text{ (right flank)}$$
$$y = 4 \cdot 307 \times 10^{-4} \, x^{1 \cdot 629} \text{ (left flank)}$$

H. Svensson (1959) found that an exponent, $n \approx 2$ ($n = 2$ represents a regular parabola) fitted certain glacial valleys in Sweden. W. L. Graf (1970) has given values of n between $1 \cdot 6$ and $2 \cdot 0$ for curves fitted to glaciated valleys in the U.S.A. The much smaller pre-exponent constant for the Antarctic case reflects its greater width.

(2) The long profile of the major trough south-west of the Otway Massif extends for at least 100 km and has an average gradient of $0 \cdot 9°$ to $1 \cdot 0°$. Deglaciated valleys of similar dimensions in the northern hemisphere possess similar gradients (for example, Yosemite: length 140 km; mean gradient $1 \cdot 1°$).

(3) The regional location of these valleys indicates that they originate in highland zones of the Transantarctic Mountains known to have been the source of early local glaciation (J. H. Mercer, 1968a—Reedy Glacier area; Mercer, 1972—Beardmore region).

The existence of inland-oriented sub-glacial valley networks has several implications. The valleys, whatever their origin (fluvial or glacial), must be related to a drainage system older than the continental ice sheet in East Antarctica. If they are glacial, and more detailed reduction by deconvolution techniques (C. H. Harrison, 1970) may further assist in determining the character and origin of these features, then they may be related to an early, mountain phase of glacierization. In such circumstances, temperate valley glaciers and small local ice caps with basal ice temperatures at the pressure melting point and with abundant free-water in the melt season, would have been capable of substantial erosion. Such an epoch of intense local glacial erosion has been suggested elsewhere in the Transantarctic Mountains. G. H. Denton *et al.* (1969, 1970) consider that the ice-free areas of southern Victoria Land were actively eroded by temperate, wet-based glaciers more than 4 million years ago. Mercer (1968a) found that by early Pleistocene times most of the erosive sequences in the Wisconsin Range had ended. Further investigations by Mercer (1972) in the Beardmore Glacier area indicate a similar early phase of intense glacial erosion.

The continental ice sheet developed from the lengthening and thickening of these inland-flowing glaciers. Once the growing ice sheet had developed sufficiently steep surface gradients the ice-flow pattern would have been drastically modified, the inland-trending valleys being submerged by the seaward-moving ice sheet. A change in the thermal régime of the ice cap must have accompanied this accumulation, basal ice temperatures falling progressively to below the pressure melting point.

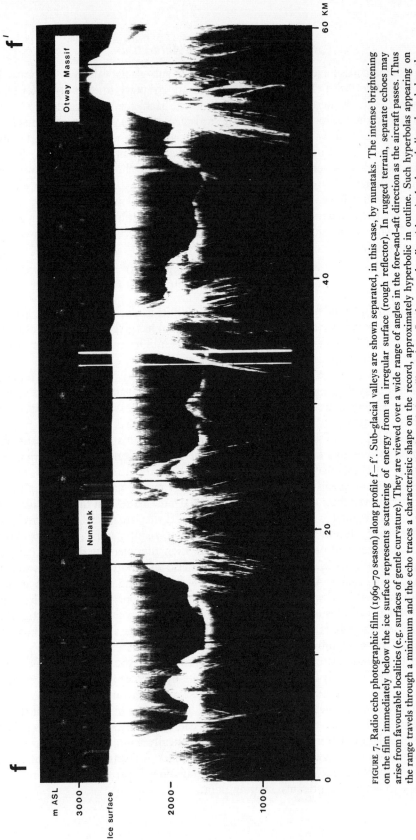

FIGURE 7. Radio echo photographic film (1969–70 season) along profile f–f'. Sub-glacial valleys are shown separated, in this case, by nunataks. The intense brightening on the film immediately below the ice surface represents scattering of energy from an irregular surface (rough reflector). In rugged terrain, separate echoes may arise from favourable localities (e.g. surfaces of gentle curvature). They are viewed over a wide range of angles in the fore-and-aft direction as the aircraft passes. Thus the range travels through a minimum and the echo traces a characteristic shape on the record, approximately hyperbolic in outline. Such hyperbolas appearing on the records do not, however, represent realistic slopes or surfaces, especially along their trailing edges. In the central valley (above), the hyperbolic echoes below the surface and near the nunatak are probably side-echoes generated by other surface peaks, but at a greater range and to one side of the flight-path

A corollary to such a sequence of events would be that the early erosional relief was little modified during the later stages of glacierization. Indeed present-day glaciological investigations (C. R. Wilson and Crary, 1961; Bentley, 1962; M. B. Giovinetto *et al.*, 1966; E. S. Robinson, 1966) show that ice flow from the plateau to the Ross Ice Shelf is small and that between the South Pole and the Queen Maud Mountains calculated basal shear stresses are low (0·2 to 0·4 × 10⁵ N m⁻²: from W. F. Budd *et al.*, 1971) and calculated basal ice temperatures approximately −10° to −20°C (ibid.). This confirms the view that the cold-based and localized ice flow in this region is unlikely to have achieved much effective erosion.

A further implication of radio echo findings is that there must have been a highland zone substantial enough to develop drainage systems and provide local centres for permanent snow accumulation (i.e., a proto-Transantarctic Mountain belt adjacent to the open water of the Ross Sea). In accordance with recent estimates for the initiation of glaciation in Antarctica (J. D. Hays and N. D. Opdyke, 1967; S. V. Margolis and J. P. Kennett, 1971; R. H. Rutford *et al.*, 1968; W. E. LeMasurier, 1972) such a highland must date from at least the Mid-Tertiary. Although a proto-Transantarctic Mountain belt has been implied in previous investigations (C. Bull *et al.*, 1962; Mercer, 1968a, 1968b; P. E. Calkin and R. L. Nichols, 1972) the new evidence of plateauward drainage pre-dating the ice sheet and effecting considerable erosion, strengthens the argument against the alternative chronology proposed by G. W. Grindley (1967). He suggests that in pre-glacial times the Transantarctic Mountains did not exist as a mountain barrier but as a 'subdued plateau landscape' (p. 562) over which the interior continental ice subsequently spread cutting the First Glacial Surface in the Miller Range at the head of Nimrod Glacier. The present elevated character of the mountains is attributed by Grindley to a response in the mantle to ice loading in the interior of East Antarctica which caused the marginal zone of the continent to rise. Mercer (1972) has shown that there has been some tectonic dislocation to early till sequences which indicates that glaciation probably commenced upon a surface undergoing the final stages of epeirogenic uplift. Whether in later stages movement was continued in response to ice loading is unconfirmed but possible.

These tentative concepts of diastrophic and glacial interaction may thus provide further evidence of the denudational history of the Transantarctic Mountains. If this belt is essentially a Mid-Tertiary relief feature and was rapidly submerged beneath an expanding ice sheet, then the primary relief elements of the mountains are tectonic and glacial. Any pre-glacial phase of sub-aerial erosion must in consequence have been extremely restricted or even absent (a hypothesis suggested by David and Priestley in 1914 and reiterated more recently by B. M. Gunn and G. Warren, 1962). Further investigations based on sub-glacial data may help to resolve these fundamental problems.

CONCLUSIONS

(1) Airborne radio echo sounding techniques now provide a rapid method of continuously profiling the ice-bedrock interface over large areas of Antarctica. (2) Radio echo soundings have indicated that the sub-glacial structural configuration inland of the Transantarctic Mountains is considerably more complex than previously envisaged. Steep scarps are found in some localities at the transition between the mountainous province and the lowland continental shield. In other areas the mountains become progressively lower and die out gradually towards the interior. Further complications are generated by discrete up-faulted

blocks or massifs found up to 600 km from the coast yet structurally part of the mountain zone.

(3) A general model for the tectonic framework of the Transantarctic Mountains is proposed in which there has been a main axis of uplift along a coastal fault producing a series of major crustal blocks which are tilted inland. Variations in tilting, faulting, igneous intrusion and subsequent erosion have produced the complex transition into East Antarctica.

(4) Evidence has been presented to suggest that Beacon Supergroup rocks may extend as far as 135° E (600 km from the Ross Sea coast at latitude 82°30′ S), confirming suggestions that the northern plainland fringe of the East Antarctic plate may comprise thin layers of sedimentary strata overlying the basement complex (A. P. Kapitsa, 1960).

(5) Radio echo soundings have indicated a pattern of sub-glacial valleys, some of which extend the bedrock network of present-day outlet glaciers in the Transantarctic Mountains. Others, however, with inland orientations are attributed to an early Mid-Tertiary mountain glacier phase. Such inland-flowing glaciers must have been, in part, responsible for the initial accumulation of the continental ice sheet but subsequent reversal of ice flow has submerged these valleys which appear to have been little modified by the restricted and cold seaward ice flow.

ACKNOWLEDGEMENTS

The programme of radio echo sounding was made possible through the effective help of the Office of Polar Programs of the U.S. National Science Foundation and the logistic support of the U.S. Navy Air Development Squadron Six. Development of radio echo sounding equipment under the direction of S. Evans and reduction of data were supported by grants from the U.K. Natural Environment Research Council. The Scott Polar Research Institute teams who carried out echo sounding programmes were: (1967–8) G. de Q. Robin, C. W. M. Swithinbank and B. M. Ewen-Smith; (1969–70) G. de Q. Robin, S. Evans, C. H. Harrison, D. L. Petrie and the author. P. E. Calkin read the draft and suggested improvements.

REFERENCES

BAILEY, J. T., S. EVANS and G. DE Q. ROBIN (1964) 'Radio echo sounding of Polar ice sheets', *Nature, Lond.* 204 (4957), 420–1

BARRETT, P. J., G. W. GRINDLEY and P. N. WEBB (1972) 'The Beacon Supergroup of East Antarctica' in *Antarctic geology and geophysics* (ed. R. J. ADIE), University of Oslo Press (in press)

BENTLEY, C. R. (1962) 'Glacial and subglacial geography of Antarctica' in *Antarctic research: the Matthew Fontaine Maury Symposium* (Washington DC; American Geophysical Union, Geophysical Monograph No. 7), 11–25

BENTLEY, C. R. (1964) 'The structure of Antarctica and its ice cover' in *Research in geophysics: solid earth and interface phenomena* (Cambridge, Mass.), Vol. 2, 335–89

BUDD, W. F., D. JENSSEN and U. RADOK (1971) 'Derived physical characteristics of the Antarctic Ice sheet', *Univ. Melbourne Meteorology Dept Publ.* No. 18, 178 pp.

BULL, C., B. C. MCKELVEY and P. N. WEBB (1962) 'Quaternary glaciations in southern Victoria Land, Antarctica', *J. Glaciol.* 4, 63–78

CALKIN, P. E., R. E. BEHLING and C. BULL (1970) 'Glacial history of Wright Valley, southern Victoria Land, Antarctica', *Antarctic J. U.S.* 5, 22–7

CALKIN, P. E. and R. L. NICHOLS (1972) 'Quaternary studies in Antarctica' in *Antarctic geology and geophysics* (ed. R. J. ADIE), University of Oslo Press (in press)

CLOUGH, J. W. and C. R. BENTLEY (1970) 'Measurements of electromagnetic wave velocity in the East Antarctic ice sheet', I.S.A.G.E. Symposium, *Int. Ass. scient. Hydrol. Publ.* No. 86, 115–28

CLOUGH, J. W. and C. R. BENTLEY (1972) 'Seismic refraction measurements of Antarctic sub-glacial structure' in *Antarctic geology and geophysics* (ed. R. J. ADIE), University of Oslo Press (in press)

CRARY, A. P. (1963) 'Results of the United States Traverses in East Antarctica 1958–1961', *I.G.Y. Glaciol. Rep.* No. 7, 144 pp.

DAVID, T. W. E. and R. E. PRIESTLEY (1914) *British Antarctic Expedition 1907–9, Report of scientific investigations. Geology* (Vol. 1) (Heinemann, London), 319 pp.

DENTON, G. H., R. L. ARMSTRONG and M. STUIVER (1969) 'Histoire glaciaire et chronologie de la région du detroit de McMurdo, sud de la terre Victoria, Antarctic: note préliminaire', *Rev. Géogr. phys. Géol. dyn.*, Ser. 2 (11) 265–78

DENTON, G. H., R. L. ARMSTRONG and M. STUIVER (1970) 'Late Cainozoic glaciation in Antarctica: the record in the McMurdo Sound region', *Antarctic J. U.S.* 5(1), 15–21

DREWRY, D. J. (1972) 'Subglacial morphology between the Transantarctic Mountains and the South Pole' in *Antarctic geology and geophysics* (ed. R. J. ADIE) University of Oslo Press (in press)

EVANS, S. (1967) 'Progress report on radio echo sounding', *Polar Rec.* 13(85), 413–20

EVANS, S. (1969) 'The VHF radio echo technique' in 'Glacier sounding symposium in the Polar regions', *Geogrl J.* 135, 547–8

EVANS S. and G. DE Q. ROBIN (1966) 'Glacier depth sounding from the air', *Nature, Lond.* 210(5039), 883–5

EVANS, S. and B. M. E. SMITH (1969) 'A radio echo equipment for depth sounding in polar ice sheets', *J. scient. Instrum.* Ser. 2 (2), 131–6

FAIRBRIDGE, R. W. (1952) 'Antarctic geology' in *The Antarctic today* (Reed, Sydney), 56–101

FORBES, M. (ed.) (1968) 'Radio echo exploration of the Antarctic ice sheet', *Polar Rec.* 14(89), 211–2

GIOVINETTO, M. B., E. S. ROBINSON and C. W. M. SWITHINBANK (1966) 'The régime of the western part of the Ross Ice Shelf drainage system', *J. Glaciol.* 6, 55–68

GRAF, W. L. (1970) 'The geomorphology of the glacial valley cross section', *Arct. alp. Res.* 2, 303–12

GRINDLEY, G. W. (1967). 'The geomorphology of the Miller Range, Transantarctic Mountains, with notes on the glacial history and neotectonics of East Antarctica', *N. Z. J. Geol. Geophys.* 10, 557–98

GUNN, B. M. and G. WARREN (1962) 'Geology of Victoria Land between the Mawson and Mulock Glaciers, Antarctica', *Transantarctic Expedition 1955–58 scient. Rep.* No. 11 (Geology) (London), 157 pp.

HARRISON, C. H. (1970) 'Reconstruction of sub-glacial relief from radio echo sounding records', *Geophysics* 35, 1099–115

HAYS, J. D. and N. D. OPDYKE (1967) 'Antarctic radiolaria, magnetic reversals and climatic change', *Science, N.Y.* 158 (3804), 1001–11

KAPITSA, A. P. (1960) 'Novye dannye o moshnosti lednikovogo pokrova tsentral'nykh rayonov Antarktidy (New data on the ice thickness cover of the central region of Antarctica)', *Inf. Byull. sov. antarkt. Eksped.* no. 19, 10–14

LEMASURIER, W. E. (1972) 'Cenozoic volcanic sequence in Marie Byrd Land and its bearing on pre-Pleistocene glaciation in Antarctica' in *Antarctic geology and geophysics* (ed. R. J. ADIE) University of Oslo Press (in press)

McGREGOR, V. R. (1963) 'Glacial or structural benches?', *J. Glaciol.* 4, 494–5

McGREGOR, V. R. and A. F. WADE (1969) 'Geology of the Western Queen Maud Mountains' in *Geologic maps of Antarctica* (Am. Geogr. Soc.) folio No. 12, sheet 16

MARGOLIS, S. V. and J. P. KENNETT (1971) 'Antarctic glaciation during the Tertiary recorded in sub-Antarctic deep sea cores', *Science, N.Y.* 170 (3962), 1085–7

MERCER, J. H. (1968a) Glacial geology of the Reedy Glacier area, Antarctica', *Bull. geol. Soc. Am.* 79, 471–85

MERCER, J. H. (1968b) 'The discontinuous glacio-eustatic fall in Tertiary sea level' in 'Tertiary sea level fluctuations', *Palaeogeogr. Palaeoclimatol. Palaeoecol.* 5, 77–85

MERCER, J. H. (1972) 'Some observations on the glacial geology of the Beardmore Glacier area, Antarctica' in *Antarctic geology and geophysics* (ed. R. J. ADIE) University of Oslo Press (in press)

NICHOLS, R. L. (1966) 'Geomorphology of Antarctica', *Am. geophys. Un. Antarctic Res. Ser.* 8, 1–46

NICHOLS, R. L. (1969) 'Geomorphic features of Antarctica' in *Geologic maps of Antarctica* (Am. Geogr. Soc.) folio No. 12, 2–6

PATERSON, W. S. B. (1969) *The physics of glaciers* (London)

ROBIN, G. DE Q. (1958) 'Seismic shooting and related investigations', *Norwegian-British-Swedish Expedition 1949–52: scientific results*, vol. 5: 'Glaciology III' (Oslo, Norsk Polarinstitut)

ROBIN, G. DE Q. (1972) 'Radio echo sounding applied to the investigation of the ice thickness and sub-ice relief of Antarctica' in *Antarctic geology and geophysics* (ed. R. J. ADIE) University of Oslo Press (in press)

ROBIN, G. DE Q., S. EVANS and J. T. BAILEY (1969) 'Interpretation of radio echo sounding in polar ice sheets', *Phil. Trans R. Soc.* Ser. A, 265 (1165), 437–505

ROBIN, G. DE Q., S. EVANS, D. J. DREWRY, C. H. HARRISON and D. L. PETRIE (1970) 'Radio echo sounding of the Antarctic ice sheet', *Antarctic J. U.S.* 5, 229–32

ROBIN, G. DE Q., C. W. M. SWITHINBANK and B. M. E. SMITH (1970) 'Radio echo exploration of the Antarctic ice sheet' I.S.A.G.E. Symposium, *Int. Ass. scient. Hydrol. Publ.* No. 86, 97–115

ROBINSON, E. S. (1962) 'Results of geophysical studies on the McMurdo to South Pole Traverse', *Res. Rep. Ser., Dept. Geol. Geophys. Polar Res., Univ. of Wisconsin*, No. 62–6, 49 pp.

ROBINSON, E. S. (1964) 'Some aspects of sub-glacial geology and glacial mechanics between the South Pole and the Horlick Mountains', *Res. Rep. Ser. Dept. Geol. Geophys. Polar Res., Univ. of Wisconsin*, No. 64–7, 88 pp.

ROBINSON, E. S. (1966) 'On the relationship of ice-surface topography to bed topography on the South Polar Plateau', *J. Glaciol.* 6, 43–54

RUTFORD, R. H., C. CRADDOCK and T. W. BASTIEN (1968) 'Late Tertiary glaciation and sea level changes in Antarctica' in 'Tertiary sea level fluctuations', *Palaeogeogr. Palaeoclimatol. Palaeoecol.* 5, 15–39

SVENSSON, H. (1959) 'Is the cross section of a glacial valley a parabola?', *J. Glaciol.* 3, 362–3

TAYLOR, G. T. (1930) *Antarctic research and adventure* (New York)

WILSON, C. R. and A. P. CRARY (1961) 'Ice movement studies on the Skelton Glacier', *J. Glaciol.* 3, 873–8

RÉSUMÉ. *La contribution des sondages par radio à l'investigation des tectoniques du cénozoique et de la glaciation en Antarctique.* Le Scott Polar Research Institute et la U.S. National Science Foundation ont entrepris un sondage par radio continuel de la ligne de démarcation entre la glace et le socle rocheux en Antarctique depuis une station aérienne pendant l'été austral de 1967-68 et celui de 1969-70. On a utilisé un système de sondage radio à 35 MHz ayant une résolution de 10m dans la glace. Ce système est beaucoup plus rapide et probablement plus exact dans la glace polaire profonde que les déterminations d'épaisseur par sondages seismiques. La technique de profil a donné de nouvelles preuves de la récente histoire diastrophique et glaciaire de l'Antarctique. Les sondages par radio le long du côté intérieur des Montagnes Transantarctiques depuis Terre Victoria et vers le sud jusqu'aux montagnes de la Reine Maud confirment de récentes investigations géologiques sur la surface, et indiquent un motif complexe de blocs causés par des failles et inclinés de façons différentes. Quelques-uns de ces blocs, comprenant probablement des roches du « Beacon Supergroup », s'étendent plus de 600 km depuis la côte de la mer de Ross. Des variations dans l'amplitude d'inclinaison, et des failles secondaires longitudinales et transversales à l'intérieur des blocs, combinées avec de probables différences spatiales dans l'érosion, ont produit une zone transitoire compliquée entre la ceinture montagneuse epicratonique et les plaines du bouclier du craton de l'antarctique orientale. Sur le côté intérieur des montagnes, des vallées probablement d'origine glaciaire, mais maintenant gisant sous la glace, indiquent que des glaciers ont descendu les deux flancs des Montagnes transantarctiques. On considère que ces vallées, et aussi celles qui pénètrent les montagnes aujourd'hui, sont dues à la glaciation locale des montagnes pendant le mi-cénozoique, et sont la conséquence du soulèvement initial de la ceinture intérieure des Montagnes Transantarctiques. Au commencement de la grande glaciation continentale, le secteur des basses terres intérieures fut submergé par entassement progressif de la glace dans l'intérieur. Le système de drainage actuel dans l'Inlandsis oriental n'est conséquemment pas en accord avec la plupart du relief sub-glaciaire. Il est possible qu'il se soit produit de petites modifications dans cette topographie enselevie dues à la glace à base froide plus récente.

FIG. 1. On peut voir le système 35MHz de sondage par radio et par air du Scott Polar Research Institute sur l'illustration ci-dessus. L'énergie électromagnétique rayonne vers le bas à partir d'antennes montées dans la queue d'un avion U.S. Navy Hercules C-130B. Le diagramme polaire indique une largeur de rayon de 25° de l'avant à l'arrière et de 10° transversalement. Des retours d'écho sont obtenus à partir de la ligne de démarcation entre des moyens de permitivité contrastés : glace/air; glace/roche ou glace/eau

FIG. 2. Un arrangement « A-scope » indique les traits saillants pendant la mesure de l'épaisseur de la glace. Le système S.P.R.I. Mark IV a un délai de 80 μs entre les pouls transmis. On peut voir des réflections retournant des surfaces glaciaires ou sous-glaciaires. La traine irrégulière de l'écho est causée par le dispersement de l'énergie sur des surfaces réflectives qui ne sont pas lisses.

FIG. 3. Des éléments morphostructurels sous-glaciaires entre les Montagnes Transantarctiques du su det le plateau du pôle sud. Le terrain sous-glaciaire au dessus de 1000m au dessus du niveau de la mer est indiqué et on peut voir la direction des vallées sous-glaciaires. Les langues glaciaires externes numerotées sont: 1) glacier Reedy, 2) glacier Robert Scott, 3) glacier Amundsen, 4) glacier Liv, 5) glacier Shackleton, 6) glacier Beardmore, 7) glacier Lennox-King, 8) glacier Marsh, 9) glacier Nimrod. W = montagnes Wisconsin, D = D'Angelo Bluff, O = Otway Massif

FIG. 4 & 5. Des profils obtenus par sondage par radio continuel entre le plateau de l'Antarctique oriental et les Montagnes transantarctiques. L'emplacement des profils est montré sur la Fig. 3

FIG. 6. Enrégistrement photographique filmé de sondage par radio de la topographie qui se développe probablement sur les roches du « Beacon Supergroup » dans l'Antarctique oriental et caracterisée par des surfaces horizontales et des escarpements raides. Le film, réalisé pendant les opérations de 1967-68, est imprimé avec l'échelle horizontale compressée et l'échelle verticale élargie. On y montre des échos retournant de la ligne de démarcation entre la glace et le socle rocheux

FIG. 7. Film photographique de sondage par radio (saison 1969-70) le long du profil (f-f'). Des vallées sous la glace se montrent séparés, dans ce cas, par des nunataks. L'éclairement intense du film immédiatement sous la surface de la glace représente l'éparpillement de l'énergie à partir d'une surface irrégulière (une réflecteur qui n'est pas lisse). Sur un terrain rugeux, des échos séparés peuvent provenir d'endroits favorables (ex. des surfaces doucement courbées). On peut les observer sous une grande variété d'angles de l'avant à l'arrière au moment où passe l'avion, et l'écho trace une forme caractéristique sur l'enrégistrement, approximativement hyperbolique. De telles hyperboles apparaissant sur les enrégistrements ne représentent pas des pentes ou des surfaces réalistes, surtout vers l'arrière. Dans la vallée centrale (en dessus) les échos hyperboliques, sous la surface, près du nunatak, sont probablement des échos secondaires, engendrés par des pics d'une autre surface, mais avec une plus grande ampleur, en dehors du chemin de vol

ZUSAMMENFASSUNG. *Der Beitrag des Radio-Echolotungens zu der Erforschung des Känozoikum-Tektoniks und der Vergletscherung in der Antarktis.* Das Scott Polar Research Institute und die U.S. National Science Foundation haben während den südlichen Sommern 1967-68 und 1969-70 ein im Flugzeug eingebautes Radio-Echolotungssystem verwendet, um ununterbrochene Profile der Grenze zwischen Eis und Felsgrund in der Antarktis zu bekommen. Es wird ein 35 MHz Radio-Echolot mit Impuls-Amplitudenmodulation und mit Auflösung von 10m in Eis verwendet. In dem tiefen Polareis ist dieses System sehr viel schneller, und wohl auch genauer, als seismische Eisdickemessungen. Diese Profilierenstechnik hat neue Beweise über die neuzeitliche diastrophische und glaziale Geschichte der Antarktis verschafft. Radio-Echolotungen, welche die inländische Seite des Transantarktischen Gebirges entlang, südwärts von Victoria-Land bis zu dem Königin-Maud-Gebirge, durchgeführt worden sind, bestätigen vor kurzem an der Oberfläche unternommene geologische Untersuchungen, und deuten auf eine komplexe Struktur von verschieden geneigten verschobenen Schollen hin. Einige dieser Schollen, welche vielleicht Felsen der ,Beacon-Supergroup' einschliessen, reichen bis 600km von der Küste des Ross-Meeres. Variationen in der Grösse der Neigungen und Sekundärverwerfungen in longitudinaler und transversaler Richtung innerhalb der Schollen, zusammen mit wahrscheinlichen räumlichen Unterschieden in der Erosion, haben eine verwickelte Übergangszone zwischen der epikrationischen Gebirgskette und dem flachen Schild des ostantarktischen Kratons verursacht. An der inländischen Seite der Berge weisen Täler von wahrscheinlich glazialer Herkunft, welche nun aber unter der Eiskalotte liegen, darauf hin, dass sich Gletscher vormals an beiden Seiten des Transantarktischen Gebirges weiterbewegten. Man nimmt an, dass diese Vertiefungen, zusammen mit denen die sich gegenwärtig auf den Bergen befinden, auf lokale Gebirgsvergletscherung während dem mittleren Känozoikum zurückzuführen sind, und der anfänglichen Erhebung des Transantarktischen Gebirges nachfolgten. Am Anfang der grossen kontinentalen Vergletscherung wurde der niedrigere inländische Sektor von progressiver Aufhäufung von Eis im Inneren überschwemmt. Das gegenwärtige Eisentleerungssystem des Inlandeises der Ostantarktis stimmt daher mit der Struktur des unterglazialen Reliefs im allgemeinen nicht überein. Diese unterglaziale Topographie ist vielleicht von neuerem kaltbasiertem Eis ein wenig verändert worden.

ABB. 1. Das im Flugzeug eingebaute Scott Polar Research Institute 35MHz Radio-Echolotungssystem wird hier gezeigt. Elektromagnetische Energie wird von Antennen im Heck eines U.S. Navy Hercules C-130B Flugzeugs nach unten ausgestrahlt. Das polare Diagramm zeigt eine Strahlenbreite von 25° in Kiellinie und 10° in transversaler Richtung. Echos werden an Grenzflächen zwischen Medien von kontrastierender Durchlässigkeit (Eis/Luft; Eis/Fels oder Eis/Wasser) zurückgeworfen

ABB. 2. Eine "A-Scope" Ausstellung zeigt hervorragende Merkmale im Laufe der Eisdickemessungen. Das S.P.R.I. Mark IV System hat eine Verzögerung von 80 μs zwischen Senderpulsen. Zurückstrahlungen von der Oberfläche vom Eis oder von unterglazialen Flächen werden gezeigt. Der unregelmässige Echoschweif resultiert aus Streuung der Energie über unebenen Reflexionsflächen

ABB. 3. Unterglaziale morphostrukturelle Elemente zwischen dem südlichen Transantarktischen Gebirge und dem südpolaren Plateau. Unterglaziales Terrain mehr als 1000m über dem Meeresspiegel und die allgemeine Richtung der unterglazialen Täler werden gezeigt. Die numerierten Gletscherströme sind: 1) Reedy-Gletscher, 2) Robert-Scott-Gletscher, 3) Amundsen-Gletscher, 4) Liv-Gletscher, 5) Shackleton-Gletscher, 6) Beardmore-Gletscher, 7) Lennox-King-Gletscher, 8) Marsh-Gletscher, 9) Nimrod-Gletscher. W = Wisconsin-Geb., D = D'Angelo-Bluff, O = Otway-Massif

ABB. 4 & 5. Profile, die durch ununterbrochene Radio-Echolotungen zwischen dem ostantarktischen Plateau und dem Transantarktischen Gebirge erhalten worden sind. Die Lage der Profile ist in Abb. 3 gezeigt

ABB. 6. Radio-Echo-Film der Topographie, die wohl auf Felsen der ,Beacon-Supergroup' in der Ostantarktis entstand, und welche für ihre horizontalen Oberflächen und steilen Landstufen charakteristisch ist. Der Film, der während den Forschungsarbeiten von 1967-68 entstanden ist, wurde mit verkleinertem horizontalen Massstab und mit vergrössertem vertikalen Massstab abgezogen. Er zeigt Echos, die von der Grenzfläche zwischen Eis und Felsboden zurückstrahlen

ABB. 7. Radio-Echo-Film (Saison 1969-70) dem Profil (f–f′) entlang. Man sieht, dass die unterglazialen Täler in diesem Falle von Nunataks getrennt sind. Die intensive Aufhellung auf dem Film, die unmittelbar unter der Eisoberfläche sichtbar ist, stellt Streuung der Energie von einer unebenen Reflexionsoberfläche dar. In unebenem Terrain kann man von günstigen Orten (z.B. von Oberflächen mit leichter Welligkeit) verschiedene Echos bekommen. Wenn das Flugzeug vorbeifliegt werden sie von vielen Aspekten in Kiellinie gesehen, und das Echo zeichnet auf den Film eine charakteristische Gestalt, die ungefähr hyperbolisch in der Kontur ist. Solche Hyperbeln auf dem Film bedeuten aber nicht realistische Neigungen oder Flächen, besonders an der Hinterkante. Die hyperbolischen Echos in dem mittleren Tal (oben), unter der Oberfläche und neben dem Nunatak, sind wohl Seitenechos, von anderen, aber weiteren Spitzen, an der Oberfläche und zur Seite des Flugwegs, verursacht

Volcanic record of Antarctic glacial history: implications with regard to Cenozoic sea levels

WESLEY E. LE MASURIER

Associate Professor of Geology, University of Colorado, Denver Center

ABSTRACT. The Pacific margin of Antarctica has been volcanically active for approximately 50 million years, with peaks of eruptive activity 18–25 and 6–12 million years ago, and throughout the Quaternary. Eruptions that took place beneath a thick glacial cover produced glassy pyroclastic deposits similar to those found in Icelandic table-mountains. The record of these deposits can be interpreted as follows: (1) an ice sheet at least several hundred metres thick was present during every eruptive episode from Eocene through Quaternary time, and (2) there were fluctuations in ice-surface level during the Quaternary but, concurrently, there were tectonic displacements that affected the elevations of moraines and glaciated surfaces. The volcanic record suggests a long and continuous history for the West Antarctic ice sheet, and this has been partly confirmed by the marine record. Together with the apparent lack of direct supporting evidence for Antarctic deglaciation, this makes it seem unlikely that there has been significant deglaciation of West Antarctica since Eocene time. The Antarctic ice sheet has apparently responded to sea-level changes that were not of its own making, but it may not have caused any large sea-level changes since its inception roughly 40 million years ago.

ALTHOUGH the Antarctic ice sheet contains enough water to raise sea level approximately 40 m (allowing for isostasy in the ocean basins), its precise role in the history of Cenozoic sea-level changes is virtually unknown. It seems certain that an ice sheet existed in Antarctica during Tertiary time, but there is no agreement about the time of its inception, and the meagre evidence of possible deglaciations during the Tertiary appears to be contradictory. During the Quaternary, there were evidently substantial fluctuations in the mass of the ice sheet, but there is debate about whether or not the Antarctic glacial maxima were in phase with northern hemisphere glacial stages (G. H. Denton *et al.*, 1970; S. Epstein *et al.*, 1970). Because of these uncertainties, the Antarctic contribution to eustatic sea-level changes remains largely unknown.

Volcanic rocks provide an independent approach to this problem. Volcanoes that have erupted beneath an ice sheet display a suite of textural and structural characteristics that are especially distinctive in volcanoes composed of basaltic lava. The pillow lavas, glass-rich tuff-breccias, palagonitic alteration, and the table-mountain landform and stratigraphy so produced, have been thoroughly described for the classic Icelandic localities by many workers including G. Kjartansson (1966), G. E. Sigvaldason (1968) and J. G. Jones (1969, 1970).[1] A review of these characteristics, with special reference to Antarctic deposits, has been presented by W. E. LeMasurier (in press).

The Icelandic studies have been largely concerned with demonstrating a sub-glacial (as opposed to sub-aerial) origin for the table-mountains and associated rock types, documenting their structural, stratigraphical and lithological characteristics, and pointing out that the genesis of table-mountains can be used as a model for the development of oceanic islands. With this background, the volcanic rocks of Antarctica are clearly relevant in any

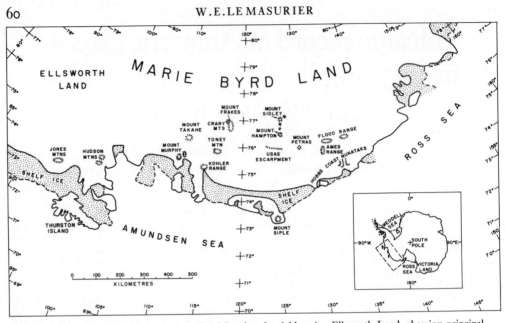

FIGURE 1. Index map of coastal Marie Byrd Land and neighbouring Ellsworth Land, showing principal
mountain ranges. The location of the map area is shown in the inset

reconstruction of glacial history. Several problems arise, however, in such a reconstruction.
(1) It is of prime importance to establish that the palagonite breccias in coastal Marie Byrd
Land are sub-glacial in origin and not submarine. (2) It is of great interest to know whether
or not the thickness of a palagonite breccia section is a measure of the thickness of ice at the
time of eruption. (3) Did volcanic sections composed of sub-aerial flows accumulate above
an ice cap, or during an ice-free interval? (4) What is the significance of trachytic hyaloclastite, in view of the fact that vitric tuffs of trachytic composition are not in themselves
diagnostic of sub-aquatic eruption, as are the basaltic hyaloclastites?

The model presented here, of uninterrupted continental glaciation in Antarctica since
Eocene time, is internally consistent with regard to the volcanic data now available, and
seems to require fewer special circumstances than any other possible interpretations. In the
following sections, the basis for this model is described and the problems referred to above
are discussed. Some aspects of Cenozoic sea levels are taken up in the concluding section.

THE NATURE OF THE VOLCANIC RECORD

Most of the geological relationships described in this paper pertain to volcanic rocks that lie
between longitude 110°W and 140°W in coastal Marie Byrd Land (Fig. 1), a region that is
comparable in size with the Japanese Islands or the British Isles. Within this region,
Cenozoic volcanic rocks can be described in terms of three geological units (Fig. 2). The
basal succession is so named because the alkaline basalt flows and hyaloclastites in this unit
form the base of the section in all the ranges where basement rock and the lower part of the
volcanic section are exposed. In general, the best exposures of the basal succession are near
the coast, where the ice sheet is thinnest. Individual basal sections may exceed 2000 m in
thickness, as described in the next section. The stratovolcanoes are composed predominantly of acid rock types such as trachyte, phonolite and alkaline rhyolite. The bases of

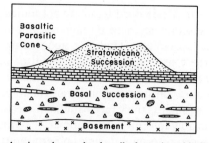

FIGURE 2. Marie Byrd Land volcanic rocks can be described stratigraphically in terms of three units, as illustrated in this diagram. Basal successions may exceed 2000 m in thickness. They are made up of sub-horizontally bedded palagonitized sideromelane tuff-breccias, of alkali basalt composition, overlain in a few places by normal (sub-aerial) basalt lava flows. The stratovolcano successions are composed largely of felsitic sub-aerial flows with initial dips of 15°–20°. Parasitic cones are composed of basalt that is chemically indistinguishable from the basal succession basalt. The Crary Mountains and Toney Mountain represent this complete sequence

stratovolcano successions are exposed only in the Flood Range, the Crary Mountains and Toney Mountain, where they overlie basal succession rocks in each case. Parasitic cones are found commonly on the flanks of the stratovolcanoes. They are composed of cinders, bombs, flows and sheets of palagonite breccia that are chemically indistinguishable from basal succession rocks. The volcanoes of Marie Byrd Land differ strikingly from circum-Pacific volcanoes outside Antarctica in the alkalinity of their lavas, the paucity of intermediate rock types, and the scarcity of pyroclastic rocks other than the palagonite breccias of the basal succession.

The most important point of comparison between the palagonite breccias of Iceland and Marie Byrd Land is their basic similarity with respect to the features that are diagnostic of sub-aquatic eruption. In each region these deposits are composed of basaltic glass fragments, marginally altered to palagonite and commonly interbedded with lenses and nodules of crystalline basalt. Similar deposits have been described from the island of Surtsey (Kjartansson, 1966) and from the deep-sea floor (E. Bonatti, 1967; Y. R. Nayudu, 1964). There seems to be little argument that these deposits represent sub-aquatic eruptions, but there appear to be no consistently recognizable differences between the products of sub-glacial and submarine volcanism. Glacial tills have been reported at the base of the hyaloclastite section in the Jones Mountains, Antarctica (R. H. Rutford *et al.*, 1968), and as interbeds in some Icelandic sections (R. W. Van Bemmelen and M. G. Rutten, 1955), but they are far too uncommon to be of diagnostic value, as may be witnessed by the fact that Kjartansson's (1943) theory of a sub-glacial origin for Icelandic table-mountains did not receive wide acceptance until roughly 20 years after it was first published. On the other hand, Bonatti's (1967) descriptions of hyaloclastites from the deep sea suggest that ferro-manganese oxide segregations and incrustations are fairly common in deep-sea deposits, and one sample contained fine-grained matrix material that was 39·9 per cent carbonate. Nayudu (personal commun.) has found pelagic fossils in some marine hyaloclastites. D. B. Clark and B. G. J. Upton (1967) describe two terrestrial occurrences of palagonite breccia, one underlain by Tertiary marine sediments in west Greenland and the other underlain by Tertiary terrestrial sediments on the east coast of Baffin Island, which they believe had a common origin and were subsequently rifted apart. They infer a submarine origin for these deposits on the basis of age (Palaeocene) and tectonic setting.

The palagonite breccias of Marie Byrd Land are not underlain by or interbedded with either glacial tills or marine sediments, and no non-volcanic material has been found, as yet, in thin sections or hand specimens. They rest upon an unusually flat erosion surface that is well exposed near the coast, but no pockets of till or marine sediment have been found on this surface, and no grooving or polishing has been observed that could not be attributed either to strong abrasion by wind and snow or to glacial erosion in the recent past. Some sections, however, such as the Crary Mountains section, are much too far inland and too elevated to be reasonably interpreted as submarine. Others are so thick (400–2000 m) that, if they are monogenetic (formed during one eruption), they would have to have been formed beyond the continental shelf, which seems very unlikely. Alternatively, repeated eruptions from the same vent area over an extended period of time, coupled with subsidence of the continental shelf at a rate commensurate with accumulation, as in a geosyncline, would have to be postulated. However, there is no other evidence for geosynclinal activity in this region during the Cenozoic, no evidence of unconformities in the hyaloclastite sections, and the vesicularity of hyaloclastite fragments suggests deep-water eruptions. These latter two points will be amplified in the next section.

Basal succession volcanism

Basal succession rocks are the most widely distributed of all Cenozoic volcanic rocks in Marie Byrd Land, in both time and space, and they are probably the most voluminous. Exposed sections overlie pre-Cenozoic basement rock at Mt Murphy, the Kohler Range, the USAS Escarpment, Mt Petras and along the Hobbs Coast, and they underlie the stratovolcanoes of the Flood Range, the Crary Mountains and Toney Mountain. They range in thickness from erosional remnants less than a few metres thick, to more than 1200

FIGURE 3. Basaltic hyaloclastites at Mt Murphy, Marie Byrd Land, showing the lenticular, sub-horizontal stratification that is common to these deposits. Thickness of the section shown is approximately 800 m; the entire section at Mt Murphy is about 2000 m. Dark lenses are crystalline basalt; light layers are palagonitized sideromelane tuff. Sample no. 3 (Table I) is from the top of the ridge in the middle ground

m in the Crary Mountains and over 2000 m at Mt Murphy (Fig. 3). Seismic evidence suggests that the basal succession underlying Toney Mountain may exceed 4000 m in thickness (C. R. Bentley and J. W. Clough, 1972). The ages of individual basal sections in Marie Byrd Land range from Eocene through Pleistocene, spanning an interval that appears, from data now available, to be at least four times the length of stratovolcano activity (Table I). Because of these characteristics, most of the record of glacial history in Marie Byrd Land has been found in basal succession deposits.

Most basal successions are composed entirely of palagonite breccia. This is one of the most striking volcanic features in the entire region. The only thick sub-aerial section of basalt is found at the west end of Toney Mountain, where the upper 200 m of the very thick section noted above is exposed. Field evidence suggests, however, that palagonite breccias underlie these sub-aerial lavas, as noted in Table I, no. 8. Two rather ambiguous outcrops of crystalline basalt, each no more than a metre thick, are found among the Hobbs Coast nunataks and in the Kohler Range (see notes, Table I, nos. 9 & 15).

Each of the remaining sections is composed of palagonite breccia, with interbedded lenses and nodules of crystalline basalt. The coloration of the breccia and the size ranges of the lenses and nodules in these sections are, to a certain extent, distinctive of individual sections. Because the composition of lava in all sections is the same, the variations in colour, texture, and internal structure from one section to another suggest that each section was produced by a relatively brief, voluminous outpouring of lava from a single vent area, the variations presumably being related to differences in the rate and violence of eruption and the depth of the volcano below the ice surface at the time of eruption. For example, the entire hyaloclastite section in the northern Crary Mountains is redder than any of the others observed in Marie Byrd Land, because of more extensive oxidation. Other sections are conspicuously yellow, because of fine grain size and extensive palagonitization, and still others are black, because of a lack of either kind of alteration. At Mt Murphy and the Crary Mountains, the crystalline lenses and nodules are much larger than those in other sections. Because these differences persist throughout the entire thickness of each section, one suspects that there are no significant unconformities, in the sense of an unconformity representing a significant period of non-deposition. Furthermore, it appears from examination of the accessible crystalline lenses and nodules that they do not have glassy selvages. This suggests that they were insulated from water by very rapid accumulation of the enclosing hyaloclastite.

The vesicularity of basaltic glass fragments suggests, further, that the basal succession hyaloclastites were erupted at depths commensurate with the thicknesses of individual sections, which seems to favour a monogenetic origin for each section rather than the 'geosynclinal' alternative mentioned earlier. J. G. Moore (1970) has shown that the vesicularity of pillow lavas changes with lava composition (tholeiitic basalt being less vesicular than alkali basalt) and with depth of eruption. A step toward extending this approach to Marie Byrd Land hyaloclastites and related glacial problems is being made by comparing the vesicularity of basal succession hyaloclastites with compositionally equivalent hyaloclastites that have been found associated with the parasitic cones. The modal analyses so far completed suggest that parasitic cone hyaloclastites are five to ten times more vesicular than hyaloclastites from the lower portions of thick basal sections, and the latter are comparable in vesicularity with alkali basalt pillows described by Moore from depths of 2500 to 3000 m. Even if the comparison is only valid in terms of order of magnitude, these are the kinds of

TABLE I

Geochronology of Cenozoic volcanic rocks in Marie Byrd Land and Ellsworth Land, West Antarctica

Sample location [Refer to Fig. 1]	Stratigraphical relationships	Age (× 10⁶ yrs)*	Series†
1. Mt Takahe (76°00′S, 112°00′W)	Sub-glacial stratovolcano	<0·240	*Pleistocene*
2. Toney Mountain (75°30′S, 116°00′W)	Supraglacial stratovolcano	0·500 (±0.2)	
3. Mt Murphy (75°15′S, 111°00′W)	Middle of a 2000+ m sub-glacial basalt section that rests on exposed basement	0·820 (±0·14)	
			——2·5 m.y.
4. Mt Bursey (76°00′S, 132°00′W)	Supraglacial stratovolcano	3·8	*Pliocene*
5. Mt Sidley (77°00′S, 126°00′W)	Supraglacial stratovolcano	6·2	
6. Shibuya Peak (75°10′S, 133°45′W)	Upper part of sub-glacial basalt section roughly 100–200 m thick, that rests on exposed basement	6·6 (±0·7)	
			——7 m.y.
7. Mt Steere (76°30′S, 118°00′W)	Lowest exposed part of 1200 m sub-glacial basalt section. No basement exposed	7·0 (±1·1)	*Miocene*
8. Toney Mountain (75°30′S, 115°00′W)	Lowest exposed part of a roughly 200 m sub-aerial basalt section. See additional note below	9·0 (±1·0)	
9. Leister Peak (75°00′S, 114°00′W)	Thin lava flow(?) on basement. See additional note below	9·8 (±1·7)	
10. Jones Mountains (73°30′S, 94°20′W)	Several different sub-glacial basalt sections	7 to 10	
11. Mt Andrus (75°45′S, 132°00′W)	Supraglacial stratovolcano	10·8 (±0·5)	
12. Mt Aldaz (76°00′S, 124°30′W)	Base of 100 m sub-glacial basalt section that overlies basement	19·4 (±1·5)	
13. Hudson Mountains (74°30′S, 100°00′W)	Sub-glacial basalt section	20 (±4)	
14. Mt Petras (75°45′S, 128°30′W)	Thin blanket of sub-glacial basalt that overlies basement rock	22·2 (±1·6)	
15. Bowyer Butte (75°00′S, 135°00′W)	Crystalline basalt sheet resting on basement See note below	23·2 (±2·1)	
			——26 m.y.
16. USAS Escarpment (75°50′S, 125°30′W)	Poorly exposed sub-glacial tuff-breccia	31·3 (±2·0)	*Oligocene*
			——38 m.y.
17. Turtle Peak (75°40′S, 111°30′W)	Base of 300–400 m sub-glacial basalt section. No basement exposed	42 (±9)	*Eocene*
18. Dorrel Rock (75°40′S, 111°30′W)	Sub-volcanic pluton of alkaline gabbro. Joints perpendicular to flow structure are occupied by aegerine syenite dikes	53·1 (±4·2)	
			——54 m.y.

* Sample description and references:

1. Aegerine syenite cognate inclusion dated by whole rock K-Ar, field 65c (LeMasurier, 1972)
2. Felsite dated by whole rock K-Ar, field 76B (LeMasurier, 1972)
3. Holocrystalline basalt dated by whole rock K-Ar, field 62A (LeMasurier, 1972)
4. Basalt dated by whole rock K-Ar (Gonzalez-F., 1971; Gonzalez-F. and Vergera, 1972)
5. Porphyritic felsite boulder dated by K-Ar (Doumani, 1963)
6. Holocrystalline basalt dated by whole rock K-Ar, field 6c (LeMasurier, 1972)

7. Holocrystalline basalt dated by whole rock K-Ar, field 73 (LeMasurier, 1972)
8. Holocrystalline basalt dated by whole rock K-Ar, field 80A (LeMasurier, 1972). Base of basalt section at this locality, determined by seismic methods, is approximately 3000 m below sea level (Bentley and Clough, 1971); total thickness is therefore about 4500m. Nearby exposure of tuff-breccia suggests that sub-glacial deposits underlie the sub-aerial basalts just beneath the present ice-surface level (LeMasurier, 1972)
9. Holocrystalline basalt dated by whole rock K-Ar, field 84 (LeMasurier, 1972). Lack of flow unit structure (e.g. glassy, vesicular crust and brecciated base), and existence of nearby tuff-breccia section, suggest that the crystalline basalt sheet at this locality is an erosional remnant of a tuff-breccia section
10. Age is derived from ten basalt samples dated by whole rock K-Ar (Rutford *et al.*, 1972). Up to 500 m of basaltic tuff-breccia overlies basement rock, at several localities, in the Jones Mountains (Rutford *et al.*, 1968)
11. Three felsite samples. Age determined by Rb/Sr isochron (Halpern, 1970)
12. Holocrystalline basalt dated by whole rock K-Ar, field 56b (LeMasurier, 1972)
13. Basalt dated by whole rock K-Ar (T. S. Laudon, unpublished manuscript)
14. Holocrystalline basalt dated by whole rock K-Ar, field 13b (LeMasurier, 1972)
15. Holocrystalline basalt dated by whole rock K-Ar, field 5 (LeMasurier, 1972). May be erosional remnant of sub-glacial tuff-breccia section
16. Holocrystalline basalt dated by whole rock K-Ar, field 58b (LeMasurier, 1972)
17. Holocrystalline basalt dated by whole rock K-Ar, field 61B (LeMasurier, 1972)
18. Gabbro dated by whole rock K-Ar, field 60A (LeMasurier, 1972)

† W. B. HARLAND *et al.* (1964)

results one would expect if basal succession volcanoes were monogenetic, formed at the base of a continental ice sheet, and if the parasitic cones were erupted beneath mountain glaciers or firn, on the flanks of supraglacial stratovolcanoes.

To summarize, there are two fundamental and interrelated questions that concern the glacial history recorded in the basal successions of Marie Byrd Land: (1) are individual basal succession volcanoes monogenetic, or do they represent successive accumulations and rejuvenation of vulcanism over relatively long intervals of time; and (2), are the palagonite breccias submarine or sub-glacial in origin? Several different approaches, none of which is independently conclusive, suggest that individual basal sections are monogenetic and were formed by eruptions beneath a thick ice cap.

If a monogenetic origin for individual basal sections is accepted, the total thickness of ice at the time of eruption would be best represented by the total thickness of palagonite breccia between basement rock and overlying sub-aerial flows. However, many of these sections do not have cappings of sub-aerial lava, and the palagonite breccia is an exceedingly friable deposit, subject to rapid mass wasting and glacial erosion even in the absence of running water. Still other sections have a sub-aerial capping, but the base of the section is not exposed. For these reasons, the thicknesses of palagonite breccia sections have been interpreted as representing minimum ice thicknesses at the time of eruption (LeMasurier, in press).

Stratovolcano successions

Stratovolcanoes make up from 50 to 100 per cent of the exposed portions of the Flood Range, Ames Range, Executive Committee Range (Mt Hampton through Mt Sidley), the Crary Mountains, Toney Mountain and Mt Takahe (Fig. 1). Fourteen different stratovolcanoes were sampled and also studied from the air. During the 1967–68 field season, the quality of exposures ranged from only two or three accessible outcrops on a few volcanoes, to a vertical cross section 1200 m high at Mt Sidley that was accessible to examination on the ground. Cliff exposures could be examined closely enough by helicopter to determine whether the rock was pyroclastic or not, and whether flow rock was basaltic or felsitic. Mt Siple could not be visited because of continually bad weather. Air photos reveal a well-preserved volcanic form with clearly defined summit caldera for this mountain, but there is no way to determine, without samples, whether it is a basal succession volcano or strato-volcano.

Field studies suggest that most of the stratovolcanoes in Marie Byrd Land are composed almost entirely of sub-aerial flows. The clearest indication of their internal structure is found at Mt Sidley, where a 1200 m section of sub-aerial flows is exposed in the north wall of the caldera. The south wall has been breached, evidently during an explosive phase of caldera collapse, and a blanket of tuffaceous ejecta perhaps 200 to 300 m thick half encircles the caldera on the south side. With apparently minor exceptions, the only other occurrences of felsitic tuffaceous rock among all the volcanoes visited are found at Mt Takahe.

Mt Takahe is an exceptional volcano because it appears to be composed almost entirely of felsitic hyaloclastite, and because its form is different from all of the other stratovolcanoes in the region (unless Mt Siple is a stratovolcano). Four well-exposed areas of outcrop were visited on the ground, and a cliff exposure was examined from the air. At each locality pyroclastic rock was the major or only rock type found. Crystalline rock was found as slabs interbedded with tuff at one locality, but was not observed as laterally continuous or superposed flows. Chemical data indicate that the rock compositions are mainly peralkaline trachyte and trachybasalt (V. H. Anderson, 1960; LeMasurier, in preparation); thus, the composition of the lava at Mt Takahe is apparently no different from that of other stratovolcanoes that are constructed of flow rock. This suggests a non-magmatic origin for the pyroclastic nature of Mt Takahe and, in this region, sub-glacial eruption is a very reasonable alternative for a volcano that is Pleistocene in age (Table I). The distinctive profile of the volcano supports this interpretation.

Figure 4 illustrates the profiles of seven volcanoes in Marie Byrd Land for which there are sufficient height data to draw profiles. Mt Murphy is the only basal succession volcano that is well enough exposed in three dimensions to be included. The profiles appear to fall into two groups with average slopes of 7° and 14°. All the sub-aerially erupted volcanoes fall into the group with steeper slopes. Mt Takahe, which is believed to be sub-glacial, falls into the group with gentle slopes. To facilitate comparisons, the profile of Mt Sidley has been projected below the ice-cap surface to approximate the same cross-sectional area as Mt Takahe. Mt Murphy also falls into the group with gentle slopes, which suggests that lava composition may not greatly influence the shapes of sub-aquatic volcanoes. One might hazard a guess from these associations that Mt Siple is composed largely of either felsitic or basaltic hyaloclastite.

Statistically, of course, the grouping of volcanoes by profile alone is weak, because the sample is small and only about half the exposed volcanoes in the region are represented. When the topographical data gathered during the recent Marie Byrd Land Survey become available, it should be possible to profile most of the remaining volcanoes. However, it is the association of a distinctive internal structure with the unusually low profile of Mt Takahe, and of Mt Murphy too, that suggests there is more validity to the grouping than statistics would imply, and the association is the basis for the interpretation just presented. A similar association has been described for volcanoes in the northern rift valleys of Ethiopia. Bonatti and H. Tazieff (1970) note that emerged submarine volcanoes composed of basaltic hyaloclastite have an unusually low ratio of height to diameter of the base, compared to neighbouring sub-aerial volcanoes.

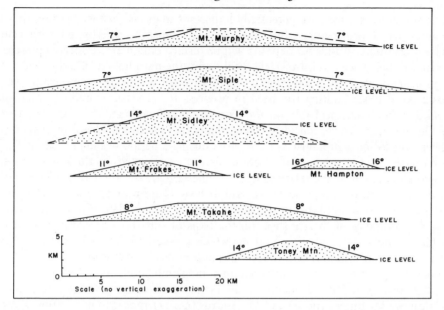

FIGURE 4. Profiles of seven Marie Byrd Land volcanoes. Mts Hampton, Sidley, Frakes, and Toney, are stratovolcanoes composed predominantly of phonolitic, trachytic, and rhyolitic flow rock. Mt Takahe is probably composed entirely of trachytic and trachybasaltic pyroclastic deposits with initial dips of $15°-20°$. Mt Murphy is a basal succession volcano composed entirely of sub-horizontally bedded palagonite breccias, as shown in Figure 3. It does not have a clearly defined summit caldera, as do the others shown. The solid line approximates to the true form of Mt Murphy omitting the irregular summit; the dashed line is for comparison with other volcanoes. All seven mountains are circular in plan, although Toney Mountain and Mt Frakes rise from an elongated chain of closely connected volcanoes. The topographical data are from U.S. Geological Survey Antarctica Sketch Map, Bakutis Coast—Byrd Land, 1:500 000, and from U.S.G.S. 1:250 000 Reconnaissance Series, Mt Sidley and Mt Hampton quadrangles

Among the stratovolcanoes, therefore, Mt Takahe is considered to be entirely subglacial in origin; all of the other stratovolcanoes (excluding Mt Siple) are almost certainly sub-aerial.

Distribution of sub-glacial and sub-aerial eruptions in time and space

Table I summarizes the ages and stratigraphical relationships of samples that are pertinent to Antarctic glacial history. The data suggest that vulcanism in both Marie Byrd Land and Ellsworth Land was episodic, with peaks of eruptive activity 18–25 and 6–12 million years ago, and during the Quaternary; consequently, the volcanic record is good for some parts of the Cenozoic and poor for others.

The palagonite breccias are the core of the volcanic record of glacial history, and the oldest of these record the existence of an ice sheet, roughly 40 million years ago, that was at least 300–400 m thick. Palagonitic tuff-breccias have been produced during every eruptive episode since that time. The well-preserved sub-glacial sections suggest ice thicknesses of more than 1200 m, 7 million years ago, and more than 2000 m during the Quaternary (Table I). Similar evidence for the existence of an ice sheet during the Tertiary has been found in the Jones Mountains and Hudson Mountains, in Ellsworth Land (Table I), and at Cape Adare in Victoria Land, East Antarctica (*U.S. Geological Survey*, 1970).

The sub-aerial sections are potentially important to glacial history because they may define intervals of deglaciation; but these deposits must be interpreted with caution. During any eruption, sub-glacial deposits will be formed only if an ice sheet is present and if the vent lies below ice level. In Marie Byrd Land, the morphology of individual ranges, the alignments of stratovolcanoes, and a variety of other geological relationships, suggest that extensional block faulting has been in progress for at least 10 million years (oldest stratovolcano date), and probably for the past 50–60 million years (LeMasurier, 1970, in press). One must, therefore, consider the possibility that a sub-aerial basal succession could have been erupted on a supraglacial horst, even during a period of intense glaciation. The stratovolcanoes, on the other hand, were evidently constructed on thick foundations of basal succession basalt, if the relationships at Toney Mountain, the Crary Mountains and the Flood Range are at all representative, and this would seem to increase the chances that these structures would build themselves above ice level early in their history. Before postulating a disappearance of the ice sheet on the basis of sub-aerial sections, therefore, one would like to find an interval of time during which sub-aerial lavas were erupted at several localities and during which there are no other indications of glaciation.

If one compares the ages of sub-aerial (including ambiguous) samples with the ages of sub-glacial samples in Table I, including the ranges of uncertainty where stated, only the stratovolcanoes Mt Bursey (no. 4) and Mt Andrus (no. 11) appear to have erupted at times when no palagonite breccias were being formed. However, Denton and others (1970) and Hamilton (in *U.S. Geological Survey*, 1970) have presented evidence that the Pliocene ice sheet on the west coast of the Ross Sea was at least as large as it is today, and perhaps larger. This suggests that Mt Bursey is supraglacial. Similarly, dated samples from Toney Mountain and Mt Sidley have yielded Pleistocene and Pliocene ages, which suggest that they too were erupted supraglacially, and Mt Hampton and Mt Berlin (Flood Range) are suspected of being late Pleistocene supraglacial volcanoes because of evidence of continuing fumarolic activity (LeMasurier and F. A. Wade, 1968). Because supraglacial stratovolcano eruptions appear to have been common, the Mt Andrus date, by itself, is an especially unconvincing piece of evidence for deglaciation 10·8 million years ago. It seems reasonable to conclude, therefore, that the volcanic data now available do not provide reliable evidence for an interval of deglaciation in Antarctica within the past 40 million years. It appears that the best hope of defining intervals of Antarctic deglaciation lies in finding non-volcanic data to fill the blanks in the volcanic record. Some possibilities are discussed in the next section.

In addition to providing evidence for a long and continuous history, the volcanic deposits of Marie Byrd Land support glaciological evidence that the Antarctic ice sheet was thicker during the Pleistocene than it is now (J. T. Hollin, 1962; Denton *et al.*, 1970). However, the relationships between sub-glacial and supraglacial sections indicate that there have been substantial tectonic displacements of Pleistocene deposits (LeMasurier, in press); the elevations of these deposits are therefore of little value in determining Pleistocene ice levels. In addition, Hollin (1962) has shown that, in coastal regions, a continental ice sheet will assume a convex cross profile if it is grounded, and a much flatter profile if the ice sheet terminates in floating ice shelves, as it does today in many parts of Antarctica including Marie Byrd Land. Therefore, if there is a lowering of sea level as there was during the northern hemisphere glaciations, the ice sheet will assume a more convex profile as it becomes grounded, and produce a thickening of ice in coastal regions that has little to do with changes in overall mass of the ice sheet. The differences between the Pleistocene ice thicknesses recorded in the volcanic deposits of coastal Marie Byrd Land, and those of today, could therefore have been produced by Hollin's mechanism rather than by a change in overall mass. The real test of this theory in Antarctica will probably not come from volcanic rocks, however, because the record is too fragmentary and too tectonically disturbed to show fluctuations of the scale of the Pleistocene interglacials, and the resolution of K-Ar dates is not good enough to place an individual section clearly within the northern hemisphere stages.

DISCUSSION

There are major gaps in the volcanic record in West Antarctica for the intervals 1–6 and 12–18 million years (m.y.) ago, and poor representation before 25 m.y. ago. Denton and others (1970) and Hamilton (in U.S. *Geological Survey*, 1970) have presented evidence for the existence of an ice sheet during the 1–6 m.y. interval, and the record in deep-sea cores supports the existence of an ice sheet in Antarctica during early Tertiary time (S. V. Margolis and J. P. Kennett, 1970). The 12–18 m.y. interval remains as an important gap in Antarctic glacial history that has yet to be filled satisfactorily. Some authors have postulated a deglaciation of Antarctica at this time, but the evidence is not compelling.

Other sources of data that might fill the 12–18 m.y. gap include a variety of palaeontological, isotopic, and oceanographical studies that bear on the related problems of Tertiary palaeotemperatures and sea levels. Most of these studies show that there were temperature fluctuations during the Tertiary and, while some maxima fall within the 12–18 m.y. interval, others occur during periods when volcanic data clearly indicate the existence of an ice sheet. Margolis and Kennett (1970), for example, note the disappearance of glacially derived quartz grains from the middle Miocene (12–15 m.y.) interval in South Pacific deep-sea cores, and they cite palaeontological and isotopic evidence for a temperature maximum in New Zealand and in sub-Antarctic waters at that time. However, the oxygen isotope data of F. H. Dorman (1966) show a temperature maximum for Australian waters in the interval 20–25 m.y. ago, with a relative low 15 m.y. ago, and C. Emiliani (1966) shows a similar relationship for middle northern and southern latitudes. Mollusc data from Victoria, Australia, coincide with Dorman's curve (E. D. Gill, 1968) and Scleractinian corals from New Zealand give evidence of a steady decline in temperature through the late Tertiary from a high in the earliest Miocene, 20–25 m.y. ago (I. W. Keyes, 1968). One could argue from these data that deglaciation, if it occurred, took place 20–25 m.y. ago, but this seems unlikely because three widely separated sub-glacial volcanic sections in Marie Byrd Land and Ellsworth Land point to the existence of an ice sheet that reached the coast during this interval. Similarly, New Zealand planktonic foraminifera show a temperature high in the late Eocene and a steady decline since that time (D. G. Jenkins, 1968), which seems to conflict with volcanic and marine evidence for Eocene glaciation in Antarctica.

It appears that Tertiary palaeotemperature fluctuations in the South Pacific do not correlate very closely with the history of the Antarctic ice sheet, and that the peaks of climatic warming vary in their timing with the locality of the study, and perhaps with the depth of water represented by each method of study. One suspects that changing patterns of oceanic circulation, and mixing of waters with different properties, must have played important roles in controlling the nature of the geological record at each locality. In this context, it seems reasonable to suggest that the opening of the Drake Passage, and consequent development of the Antarctic circumpolar current, had a very important effect on oceanic circulation in the South Pacific, and may have been responsible for the disappearance of glacially derived quartz grains from the middle Miocene intervals of some deep-sea cores.

There is little direct evidence concerning Tertiary sea levels that can be used as a constraint on estimates of Antarctic glacial history, because various models of ice-sheet history have, in fact, usually served as the major basis for estimating fluctuations in Tertiary sea level (W. F. Tanner, 1968, p. 12). Data that are independent of Antarctic history, and that could bear a direct relationship to Tertiary sea-level changes, seem to conflict with or be unrelated to the history of Antarctic glaciation. S. Gartner (1970), for example, notes a major lowering of the carbonate dissolution level during the Pliocene and suggests that this may have been related to a 100 m drop in sea level that took place during the formation of the Antarctic ice sheet. S. D. Webb and N. Tessman (1967) describe fossiliferous beach deposits in Florida which suggest that the sea approached its present level 4 to 7 million years ago. It seems unlikely, however, that any large lowering of sea level in Pliocene time was related to Antarctic glaciation, because a substantial body of evidence, in addition to that presented in this paper, indicates that the Pliocene ice cap was fully as large as it is today, or larger (H. G. Goodell et al., 1968; Denton et al., 1970; U.S. Geological Survey, 1970).

In summary, it appears that middle Miocene time (12–18 m.y. ago) is the only interval within the last 40 million years for which there is no direct evidence of glaciation in Antarctica; but a review of pertinent palaeontological, isotopic and oceanographical data shows that there is also no reliable evidence that the ice sheet disappeared during that interval. To be more speculative, it has been argued that, once the ice cap became large enough to waste principally by calving, it would be unlikely to disappear during climatic warming cycles, such as the Pleistocene interglacials (Hollin, 1962, 1970). The volcanic sections noted in Table I are all close enough to the coast (within 0–200 km) to indicate that the ice sheet was calving into the sea during each eruptive episode throughout the Cenozoic. Finally, the future probability of finding volcanic evidence for deglaciation seems small, because the only sub-aerial basal sections found during the 1967–68 Marie Byrd Land Survey and the 1968–69 Ellsworth Land Survey (F. A. Wade and K. E. LaPrade, 1969) are those noted earlier, which have provided ambiguous evidence; there are, however, several palagonitic sections that still remain to be dated.

The volcanic record thus renders it unlikely that Antarctica has been deglaciated since the Eocene and, therefore, eustatic sea-level controls outside Antarctica should be evaluated very carefully before postulating large fluctuations in the Antarctic ice sheet. For example, several recent discoveries suggest that northern hemisphere glaciations must have had an important influence on late Tertiary sea levels. W. O. Addicott (1969), O. L. Bandy et al. (1969), and Denton and Armstrong (1969) have presented a variety of data that point to the existence of glaciers in south-eastern Alaska in middle Miocene time, about 10–13 m.y. ago. Y. Herman (1970) presents evidence from Arctic Ocean cores that high latitude glaciation began prior to 6 m.y. ago, and preliminary results from Leg 12 of the Deep Sea Drilling Project suggest that glaciers began to cover the continents about 3 m.y. ago (A. S. Laughton et al., 1970). There appears to be some consistency between the implications of these data and those of Webb and Tessman (1967). The latter suggest that, in early Pliocene time (4–7 m.y. ago), the sea stood essentially at today's level, which is unusually low if one assumes that there were no glaciers on earth at that time. However, the data from the Arctic and Antarctic suggest that glaciation was about as extensive in late Miocene and early Pliocene time as it is today. Middle Pliocene time evidently saw the beginnings of continental glaciation in the northern hemisphere and, from that time on, glacio-eustatic sea-level changes were probably controlled predominantly by these ice sheets, because the Antarctic ice sheet already covered the entire continent and could not grow substantially larger until sea level was lowered.

In conclusion, the following sequence of events is proposed as a tentative and rudimentary framework of Tertiary sea-level history. (1) A major glacio-eustatic lowering of sea level took place in Eocene time, caused by the initial development of a continental ice sheet, as the Antarctic continent drifted into its present polar position (LeMasurier, in press). This lowering may be reflected in early Tertiary events such as the draining of the Mississippi Embayment in North America. (2) A second major lowering took place, beginning in early to mid-Pliocene time and culminating in the Pleistocene, that was related to the growth of northern hemisphere ice sheets.

In detail, this sequence may have been complicated by factors that are not normally considered in sea-level reconstructions. Within the past 40 million years of Cenozoic glacial history it is possible that sea levels have been significantly affected by changes in the volume of the ocean basins related to sea-floor spreading (E. M. Moores, 1970) and by changes in the shape of the geoid. Maps of the geoid presented by E. M. Gaposchkin and K. Lambeck (1970) and by D. King-Hele (1967) show a sea-level relief of 194 m and 146 m, respectively, superimposed on the basic spheroidal shape of the earth. Whether these 'highs' and 'lows' on the sea-level surface migrate, as do the continents, at rates of a few centimetres a year, or whether they simply remain stationary as the continents migrate, they could produce large and puzzling complexities in the long-term record of sea-level changes. If the glacio-eustatic factors in sea-level fluctuations can be isolated it may then become possible to recognize some of these other factors.

ACKNOWLEDGEMENTS

Field investigations connected with this study were carried out during the 1967–68 Marie Byrd Land Survey, supported by NSF Grant GA-5469, administered by the Office of Polar Programs. Laboratory studies of samples collected during the survey have been supported by NSF Grants GA-4563 and GA-21488. I am grateful to Drs. J. T. Andrews, P. W. Birkeland, W. C. Bradley and D. L. Eicher for critically reading the manuscript.

NOTE

1. The terms hyaloclastite and palagonite breccia have been used more or less interchangeably by various authors, in reference to palagonitized sideromelane tuff-breccias of both sub-glacial and submarine origin.

REFERENCES

ADDICOTT, W. O. (1969) 'Tertiary climatic change in the marginal north-eastern Pacific Ocean', *Science, N.Y.* 165, 583–5

ANDERSON, V. H. (1960) 'The petrography of some rocks from Marie Byrd Land, Antarctica', *USNC-IGY Antarct. glaciol. Data 1958–59*, Rep. 825–2, Part VII, 27 pp.

BANDY, O. L., A. E. BUTLER, and R. C. WRIGHT (1969) 'Alaskan Upper Miocene marine glacial deposits and the *Turborotalia Pachyderma* datum plane', *Science, N.Y.* 166, 607–9

BENTLEY, C. R. and J. W. CLOUGH (1972) 'Seismic refraction measurements of Antarctic subglacial structure' in *Antarctic geology and geophysics* (ed. R. J. ADIE), University of Oslo (in press)

BONATTI, E. (1967) 'Mechanism of deep sea volcanism in the South Pacific' in *Researches in geochemistry* (ed. P. H. ABELSON), vol. 2, 453–91

BONATTI, E. and H. TAZIEFF (1970) 'Exposed guyot from the Afar Rift, Ethiopia', *Science, N.Y.* 168, 1087–9

CLARK, D. B. and B. G. J. UPTON (1967) 'Tertiary basalts of Baffin Island: field relations and tectonic setting', *Can. J. Earth Sci.* 8, 248–58

DENTON, G. H. and R. L. ARMSTRONG (1969) 'Miocene-Pliocene glaciations in southern Alaska', *Am. J. Sci.* 267, 1121–42

DENTON, G. H., R. L. ARMSTRONG and M. STUIVER (1970) 'Late Cenozoic glaciation in Antarctica: the record in the McMurdo Sound region', *Antarctic J. U.S.* 5, 15–21

DORMAN, F. H. (1966) 'Australian Tertiary palaeotemperatures', *J. Geol.* 74, 49–61

DOUMANI, G. A. (1963) 'Volcanoes of the Executive Committee Range, Byrd Land' in *Antarctic geology* (ed. R. J. ADIE), Amsterdam, North Holland Publ. Co., 666–75

EMILIANI, C. (1966) 'Isotopic palaeotemperatures', *Science, N.Y.* 154, 851–7

EPSTEIN, S., R. P. SHARP and A. J. GOW (1970) 'Antarctic ice sheet: stable isotope analyses of Byrd Station cores and interhemispheric climatic implications', *Science, N.Y.* 168, 1570–2

GAPOSCHKIN, E. M. and K. LAMBECK (1970) '1969 Smithsonian standard earth (II)', *Smithson. Astrophys. Observ. Spec. Rep.* 315, 93 pp.

GARTNER, S., Jr. (1970) 'Sea-floor spreading, carbonate dissolution level, and the nature of horizon A', *Science, N.Y.* 169, 1077–9

GILL, E. D. (1968) 'Oxygen isotope palaeotemperature determinations from Victoria, Australia', *Tuatara* 16, 56–61

GONZALEZ-FERRAN, O. (1972) 'Distribution, migration and tectonic control of upper Cenozoic volcanism in West Antarctica and South America' in *Antarctic geology and geophysics* (ed. R. J. ADIE), University of Oslo (in press)

GONZALEZ-FERRAN, O. and M. VERGARA (1972) 'Post-Miocene volcanic petrographic provinces of West Antarctica and their relation with the southern Andes of South America' in *Antarctic geology and geophysics* (ed. R. J. ADIE), University of Oslo (in press)

GOODELL, H. G., N. D. WATKINS, T. T. MATHER and S. KOSTER (1968) 'The Antarctic glacial history recorded in sediments of the southern ocean, *Palaeogeogr. Palaeoclimatol. Palaeoecol.* 5, 41–62

HALPERN, M. (1970) 'Rubidium-strontium dates and Sr^{87}/Sr^{86} initial ratios of rocks from Antarctica and South America: a progress report', *Antarctic J. U.S.* 5, 159–61

HARLAND, W. B., A. G. SMITH and B. WILCOCK (eds.) (1964) 'The Phanerozoic time scale', *Q. J. geol. Soc. Lond.* 120

HERMAN, Y. (1970) 'Arctic palaeo-oceanography in late Cenozoic time', *Science, N.Y.* 169, 474–7

HOLLIN, J. T. (1962) 'On the glacial history of Antarctica', *J. Glaciol.* 4, 173–95

HOLLIN, J. T. (1970) 'Is the Antarctic ice sheet growing thicker?' in *International symposium on Antarctic glaciological exploration (ISAGE)*, Hanover, N. H. IASH Publ. 86, 363–74

JENKINS, D. G. (1968) 'Planktonic Foraminiferida as indicators of New Zealand Tertiary palaeotemperatures', *Tuatara* 16, 32–7

JONES, J. G. (1969) 'Intraglacial volcanoes of the Laugarvatn region, south-west Iceland, I', *Q. J. geol. Soc. Lond.* 124, 197–211

JONES, J. G. (1970) 'Intraglacial volcanoes of the Laugarvatn region, south-west Iceland, II', *J. Geol.* 78, 127–40

KEYES, I. W. (1968) 'Cenozoic marine temperatures indicated by the Scleractinian coral fauna of New Zealand', *Tuatara* 16, 21–5

KING-HELE, D. (1967) 'The shape of the earth', *Scient. Am.* 217, 67–76

KJARTANSSON, G. (1943) 'Jardsaga', *Arnesinga Saga* (Reykjavik), 248 pp.

KJARTANSSON, G. (1966) 'A comparison of tablemountains in Iceland and the volcanic island of Surtsey off the south coast of Iceland' (text in Icelandic with English summary), *Náttúru fræðingurinn* 36, 1–34

LAUGHTON, A. S. and W. A. BERGGREN (1970) 'Deep sea drilling project Leg 12', *Geotimes* 15 (II), 10–14

LEMASURIER, W. E. (1970) 'Tectonic environment of circum-Pacific volcanism in Marie Byrd Land, Antarctica' (abstr.) *Trans. Am. geophys. Un.* 51, 824

LEMASURIER, W. E. (1972) 'Volcanic record of Cenozoic glacial history, Marie Byrd Land' in *Antarctic geology and geophysics* (ed. R. J. ADIE), University of Oslo (in press)

LEMASURIER, W. E. and F. A. WADE (1968) 'Fumarolic activity in Marie Byrd Land, Antarctica', *Science, N.Y.* 162, 352

MARGOLIS, S. V. and J. P. KENNETT (1970) 'Antarctic glaciation during the Tertiary recorded in sub-Antarctic deep-sea cores', *Science, N.Y.* 170, 1085–7

MOORE, J. G. (1970) 'Water content of basalt erupted on the ocean floor', *Contr. Miner. Petrology* 28, 272–9

MOORES, E. M. (1970) 'Patterns of continental fragmentation and reassembly: some implications' (abstr.) *Geol. Soc. Am. Abstracts of Programs* 2, 629

NAYUDU, Y. R. (1964) 'Palagonite tuffs (hyaloclastites) and the products of post-eruptive processes', *Bull. volcan.* 27, 391–410

RUTFORD, R. H., C. CRADDOCK, R. L. ARMSTRONG and C. M. WHITE (1972) 'Tertiary glaciation in the Jones Mountains' in *Antarctic geology and geophysics* (ed. R. J. ADIE), University of Oslo (in press)

RUTFORD, R. H., C. CRADDOCK and T. W. BASTIEN (1968) 'Late Tertiary glaciation and sea-level changes in Antarctica', *Palaeogeogr. Palaeoclimatol. Palaeoecol.* 5, 15–39

SIGVALDASON, G. E. (1968) 'Structure and products of subaquatic volcanoes in Iceland', *Contr. Miner. Petrology* 18 1–16

TANNER, W. F. (1968) 'Tertiary sea-level symposium—introduction', *Palaeogeogr. Palaeoclimatol. Palaeoecol.* 5, 7–14

U.S. *Geological Survey* (1970) 'Geological Survey research, 1969', *U.S. geol. Surv. Prof. Pap.* 650-A

VAN BEMMELEN, R. W. and M. G. RUTTEN (1955) *Tablemountains of northern Iceland and related geologic notes* (E. J. Brill, Leiden), 217 pp.

WADE, F. A. and K. E. LAPRADE (1969) 'Geology of the King Peninsula, Canisteo Peninsula, and Hudson Mountains areas, Ellsworth Land, Antarctica', *Antarctic J. U.S.* 4, 92–3

WEBB, S. D. and N. TESSMAN (1967) 'Vertebrate evidence of a low sea level in the Middle Pliocene', *Science, N.Y.* 156, 379

RÉSUMÉ. *Enrégistrement volcanique de l'histoire glaciale de l'Antarctique: implications quant aux niveaux de la mer cenozoïque.* La marge Pacifique de l'Antarctique a été active volcaniquement pendant approximativement 50 millions d'années, avec apogées d'activités éruptives il y a 18 à 25 millions d'années et de 6 à 12 millions d'années, et pendant tout le Quaternaire. Les éruptions qui ont eu lieu audessous d'une épaisse couche glaciale ont produit des sédiments vitreux pyroclastiques semblables à ceux trouvés dans les plateaux Islandais. L'enrégistrement de ces sédiments peut être interprété de la suite: (1) une couche épaisse de glace (au moins plusieurs centaines de mètres) était présente pendant chaque période éruptive du temps Eocène jusque pendant le Quaternaire, et (2) il y avaient des fluctuations du niveau de la glace pendant le Quaternaire, mais simultanément il y avaient des déplacements tectoniques qui ont affecté les élévations de moraines et des surfaces striées. Les roches volcaniques suggèrent une histoire très longue et continuelle de la couche de glace de l'Antarctique de l'ouest, et ceci a été en partie confirmé par l'histoire des sediments maritimes. De ce perspectif historique, et considérant le manque d'évidence apparente dircte du dégèlement de l'Antarctique, il semble improbable qu'il y ait eu un dégèlement significatif de l'Antarctique de l'ouest depuis la période Eocène. La couche de glace de l'Antarctique a été apparemment sensible aux changements du niveau de la mer, ces changements ne se sont pas faits d'eux-mêmes, mais c'est possible qu'il ne peut pas avoir causé de grands changements au niveau de la mer depuis son commencement il y a approximativement 40 millions d'années.

FIG. 1. Index de la carte côtière de Marie Byrd voisinant Ellsworth Land et montrant les chaînes de montagnes principales. La location de la carte est montrée dans le hors-texte

FIG. 2. Les roches volcaniques du Marie Byrd Land peuvent être décrites stratigraphiquement en termes de trois éléments, comme illustrés dans ce schéma. Les successions de base peuvent surpasser 2000 m d'épaisseur. Elles sont faites de couches presqu'horizontales de hyaloclastite, d'une composition alcali basalte, recouvrant en quelques endroits par des coulées normales de laves de basalte. Les successions de couches stratovolcaniques sont composés grandement de coulées felsitiques avec des pendages initiales de 15°à 20°. Les cônes parasitiques sont composés de basalte chimiquement semblable aux basaltes de la succession de base. Les Crary Mountains et le Toney Mountain représentent cette série complète

FIG. 3. Les hyaloclastites basaltiques du Mt Murphy, Marie Byrd Land, montrent la stratification lentiforme qui est commune à ces gisements. L'épaisseur de la section montrée est approximativement de 800 m. La section entière du Mt Murphy est de 2000 m environ. Les lentilles noires sont de basalte cristalliné; les couches claires sont hyaloclastites. Échantillon 3 (Fig. 1) est du sommet de la butte au fond du motif

FIG. 4. Profils de sept volcans de Marie Byrd Land. Mt Hampton, Mt Sidley, Mt Frakes et Toney Mountain sont des stratovolcans composés de coulées d'une prédominance phonolite, trachyte, et rhyolite. Le Mt Takahe est probablement composé entièrement de hyaloclastites trachytiques et trachybasaltiques avec des pendages initiaux de 15° à 20°. Le Mt Murphy est un volcan de succession basale composé entièrement de couches presqu'horizontales de brèches palagonites, comme montré dans la Figure 3. Mt Murphy n'a pas de caldera clairement défini au sommet, comme les autres l'ont montré. La ligne solide se rapproche de la forme vrai du Mt Murphy en omettant le sommet irrégulier; les tirets sont pour comparer les autres volcans. Toutes les sept montagnes sont d'un plan circulaire, bien que Toney Mountain et Mt Frakes se lèvent d'une chaîne élongée de volcans connectés de près. Les données topographiques sont du U.S. *Geological Survey Antarctica Sketch Map, Bakutis Coast, Byrd Land, 1:500 000*, et du *USGS 1:250 000 Reconnaissance Series* Mt Sidley et Mt Hampton quadrangles

ZUSAMMENFASSUNG. *Eine vulkanische Geschichte der antarktischen Glazialzeit: Folgerungen in Bezug auf känozoische Meeresspiegelhöhen.* Im pazifischen Abschnitt der Antarktis gibt es seit ungefähr 50 Millionen Jahren vulkanische Tätigkeit, die vor 18–25 Millionen Jahren, wiederum vor 6–12 Millionen Jahren und während des ganzen Quatärs eruptive Höhepunkte erreicht hat. Ausbrüche, die unter einem dicken Gletschermantel stattfanden, erzeugten glasige, pyroklastische Ablagerungen, die den auf den isländischen Tafelbergen gefundenen Ablagerungen ähneln. Die Geschichte dieser Ablagerungen lässt sich folgendermassen auslegen: (1) Eine mindestens etliche hundert Meter dicke Eisschicht war während jedes eruptiven Zeitabschnitts vom Eozän durch das Quatär vorhanded, und (2) es gab Schwankungen im Stand der Eisoberfläche während des Quatärs, aber gleichzeitig fanden tektonische Verschiebungen statt, die sich auf die Höhe der Moränan und vergletscherten Oberflächen auswirkten.

Aus dieser vulkanischen Geschichte lässt sich ein langes und ununterbrochenes Bestehen der westantark-
tischen Glazialschicht schliessen. Diese Schlussfolgerung ist zum Teil durch die Meeresgeschichte bestätigt worden.
Gesehen aus dieser geschichtlichen Perspecktive scheint es unwahrscheinlich, dass wesentliche Entgletscherung der
Westantarktis seit dem Eozän stattgefunden hat; zumal es augenscheinlich en einschlägigem Beweis für antarktische
Entgletscherung mangelt. Anscheinend hat die antarktische Glazialschicht auf Schwankungen der Meeresspiegel-
höhe, die sie selbst nicht hervorgerufen hat, reagiert; aber sie dürfte selbst keine grossen Änderungen der Meeres-
spiegelhöhe seit ihrer Entstehung vor etwa 40 Millionen Jahren hervorgerufen haben.

ABB. 1. Indexkarte vom Kustengebiet des Marie-Byrd-Land und vom benachbarten Ellesworth-Land. Die Haupt-
gebirge sind gezeigt. Die Lage des auf der Karte dargestellten Gebiets befindet sich im Eckeinsatz

ABB. 2. Vulkanische Gesteine des Marie-Byrd-Land konnen stratigraphisch in der From von drei Einheiten dargestellt
werden, wie diese Abbildung zeigt. Die Dicke der Grundschichten kann mehr als 2000 m betragen. Sie bestehen aus
beinahe waagerecht verlaufenden, palagonitisierten, glasernen Basalttuffbrecchien von einer Alkalibasaltzusammen-
setzung, die an vereinzelten Stellen mit normalen (subäerischen) Basaltlavastromen uberlagert sind. Die stratovulkane setzen sich zum grossen Teil aus felsitischen, subäerischen Stromen mit einer ursprunglichen Neigung von
15°–20°. Die parasitaren Kegel bestehen aus Basalt, der sich vom Basalt der Grundschicht chemikalisch nicht unter-
scheiden lasst. Das Crary-Mountains und der Toney-Mountain vertreten diese vollstandige Reihenfolge

ABB. 3. Basalthyaloklastite auf dem Mt Murphy im Marie-Byrd-Land. Gezeigt ist die linsenartige, beinahe waagerechte
Schichtung, die diesen Ablagerungen eigen ist. Die Dicke des dargestellten Abschnitts misst ungefahr 800 m; der
vollstandige Querschnitt mit allen Schichten am Mt Murphy misst etwa 2000 m. Das Material der schwarzen Linsen
ist Kristallbasalt, das der hellen Schichten ist palagonitisierter, glaserner Basalttuff. Probe Nr. 3 (Tabelle I) stammt
vom Gipfel des Kammes im Mittelgrund

ABB. 4. Durchschnitte von sieben Vulkanen im Marie-Byrd-Land. Die Berge Hampton, Sidley, Frakes und Toney
sind Stratovulkane, die vorwiegend aus Phonolith-, Trachyt- und Rhyolith- Fliessgestein bestehen. Mt Takahe
durfte ganz aus pyroklastischen Trachyt- und Trachy-basalt-Ablagerungen bestehen, die eine ursprunglichen Neigung von 15°–20° aufweisen. Mt Murphy ist ein Grundschichtenvulkan, der sich lediglich aus fast waagerecht
verlaufenden Palagonitbrecchien zusammensetzt, wie Abbildung 3 darstellt. Er hat keine klar umrissene Gipfelkaldera
wie die ubrigen dargestellten Vulkane. Die ununterbrochene Linie stellt annahernd die wirkliche Gestalt von Mt
Murphy ohne den unregelmassigen Gipfel dar. Die unterbrochene Linie dient zum Vergleich mit anderen Vulkanen.
Jeder der sieben Berge hat einen kreisformigen Grundriss, obwohl sich Toney-Mountain und Mt Frakes aus einer
ausgedehnten Kette eng zusammenhangender Vulkane erheben. Topographische Daten sind aus der *U.S. Geological
Survey Antarctica Sketch Map (fortgesetzt) Bakutis Coast—Byrd Land*, 1:500 000, und aus *U.S.G.S. 1:250 000
Reconnaissance Series*, aus den Vierecken von Mt Hampton und Mt Sidley

Evidence from the South Shetland Islands towards a glacial history of West Antarctica

BRIAN S. JOHN

Lecturer in Geography, University of Durham

Revised MS received 29 October 1971

ABSTRACT. The glacial history of the McMurdo region is now moderately well known, but there is little reliable evidence on which the glacial chronology of West Antarctica may be based. J. H. Mercer has postulated that a West Antarctic ice sheet of Saalian age collapsed catastrophically during the Eemian interglacial, to be followed by renewed ice-sheet growth during the Weichselian stage. Evidence is presented from the Maxwell Bay area, central South Shetland Islands, to support this hypothesis. During an early (Saalian?) glaciation the head of Maxwell Bay was inundated by ice moving south-eastwards from an ice cap based on the shallow submarine shelf to the north of the South Shetland Islands. This ice was responsible for the streamlining of bedrock relief on Fildes Peninsula, and for the erosion of the glacial trough which is now submerged in Maxwell Bay. The ice cap wasted rapidly at the onset of an (Eemian?) warmer phase, leading to the erosion of anastomosing channel systems by sub-glacial meltwater. The altitudes of 'residual beaches' (found up to 275 m) around Maxwell Bay are possibly the result of later compensational uplift which accompanied renewed ice-sheet growth and isostatic depression in the continental interior. During the Weichselian there was a local glaciation of limited duration and intensity, in which ice from the island ice caps over-rode previously ice-free peninsulas and removed most of the pre-existing raised beaches. Flandrian ice-edge withdrawal and isostatic recovery was accompanied by the formation of raised beach-ridges below 54 m, and there was a short-lived glacier readvance which culminated at about 500 years B.P. This sequence of events may be representative of a much wider area and in particular the Antarctic Peninsula sector of West Antarctica.

IN recent years there have been great advances in the knowledge of the glacial history of Antarctica as a result of glaciological, geomorphological, oceanographical and geological investigations on an unprecedented scale. As P. E. Calkin and R. L. Nichols (in press) have pointed out, most of these advances have been made in the Ross Sea sector, and more particularly in the ice-free areas near McMurdo Sound. Here, G. H. Denton *et al.* (1969) have been able to produce a chronology for several recognizable glacial and non-glacial stages. In West Antarctica the search for a comparable record has been handicapped by the lack of extensive ice-free lowland areas where unconsolidated ancient tills or interglacial deposits might have survived. There are tillites in the Jones Mountains which indicate that the initial accumulation of the continental ice sheet occurred during the Tertiary (R. H. Rutford *et al.*, 1968), but little is known of the course of Quaternary glaciation in West Antarctica. For example, workers in the Antarctic Peninsula sector have been able to state only that the ice cover of coastal areas was once thicker and more extensive than at present (W. L. S. Fleming, 1940; G. de Q. Robin and R. J. Adie, 1964; D. L. Linton, 1964). Little reliable evidence has been obtained which would allow this 'maximum' stage of Quaternary glaciation to be dated. On the other hand, the recent thinning of the ice cover is well documented in the isostatically raised storm-beach series found in many coastal embayments (Nichols, 1947; Adie, 1964a), and in the relationships between these storm-beaches and adjacent outlet-glacier moraines (B. S. John and D. E. Sugden, 1971).

In an interesting recent paper J. H. Mercer (1968a) presented a hypothesis concerning the glacial stages of West Antarctica. Assuming a broad synchroneity of northern and southern hemisphere glacial stages and non-glacial intervals, he suggested that during the Saalian glacial stage the West Antarctic ice sheet was fully developed. Later, during the Eemian interglacial, temperatures rose until they were too high for the continued existence of ice shelves. As a result, that part of the West Antarctic ice sheet which was grounded below sea level disintegrated rapidly, causing a eustatic sea-level rise to about 6 m some 120 000 years ago. Temperatures then fell again, and the West Antarctic ice sheet of the Weichselian stage was established. This ice sheet has survived to the present day. The hypothesis of an Eemian ice-sheet collapse may conveniently be matched with J. T. Hollin's (1969; 1970a) hypothesis of Antarctic ice surges; and indeed it receives some support in the glaciological record of the Byrd Station ice-sheet core (S. Epstein *et al.*, 1970). But there is as yet no unequivocal geomorphological evidence for large-scale melting of the West Antarctic ice sheet during the last interglacial. The 'warm intervals' traced by Mercer (1968b) in the vicinity of Reedy Glacier are difficult to date, and it may be that traces of interglacial landforms and sediments within the sphere of influence of the West Antarctic ice sheet have been destroyed or obscured during the most recent stage of ice expansion. Possibly it is more worthwhile to search for such traces either on the coasts of the Antarctic Peninsula beyond the present northern limits of shelf ice (Robin and Adie, 1964), or on the sub-Antarctic islands of the Scotia Ridge. These islands have probably experienced the same succession of major climatic episodes as the West Antarctic ice sheet, but their glaciations will have been less intensive and their non-glacial intervals more prolonged. In this paper it is suggested that the Maxwell Bay area of the central South Shetland Islands may prove to be of considerable importance for the glacial geomorphology of West Antarctica; several lines of evidence appear to provide support for Mercer's hypothesis.

GEOLOGICAL AND PHYSICAL BACKGROUND

The central South Shetland Islands are situated at approximately longitude 60°W and latitude 62°S, on the southern flank of Drake Strait and some 140 km from the tip of the Antarctic Peninsula (Fig. 1). The islands of the group are essentially part of the Scotia Ridge, being composed largely of Jurassic to Tertiary volcanic and sedimentary rocks associated with lines of structural weakness (Adie, 1964a). Volcanic activity has continued through the Quaternary on a smaller scale (C. M. Barton, 1965), and in recent centuries Bridgeman Island and Deception Island have experienced eruptions (P. E. Baker *et al.*, 1969). Areas of alpine relief are limited to parts of Livingston Island and Greenwich Island, and for the most part the landscapes are dominated by low ice domes fringed at the coastline by ice cliffs. There are a few nunataks in the interiors of the islands, but the most extensive ice-free areas are Byers Peninsula (*c*. 66 km²) on Livingston Island, and Fildes Peninsula (*c*. 60 km²) and Barton Peninsula (*c*. 15 km²) on King George Island. There are many other small ice-free peninsulas, headlands and islands scattered around the coasts, while the exposed northern shores are characterized by a wide belt of skerries (O. Holtedahl, 1929). Below an altitude of 120 m on the major ice-free peninsulas the dominant landscape features are extensive erosion surfaces and ancient marine clifflines; the latter are often 30 m high. Both surfaces and clifflines are apparently pre-glacial, and have been modified to a variable extent by glaciation during the Quaternary (John and Sugden, 1971).

Maxwell Bay is a broad inlet opening south-eastward into Bransfield Strait. It is

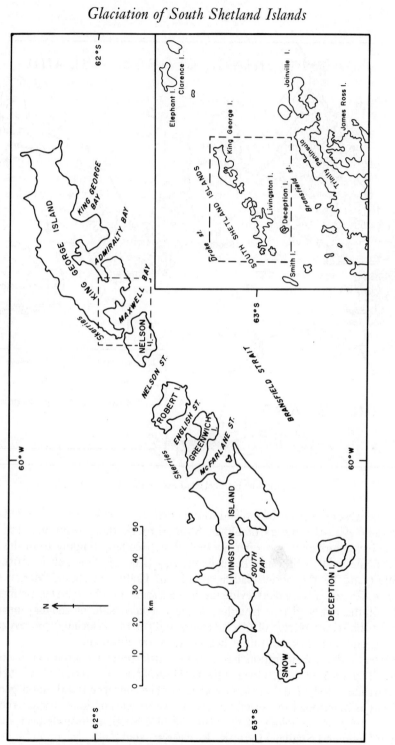

FIGURE 1. The South Shetland Islands and their situation with respect to the tip of the Antarctic Peninsula. Inset box around Maxwell Bay indicates the area covered in Figure 2

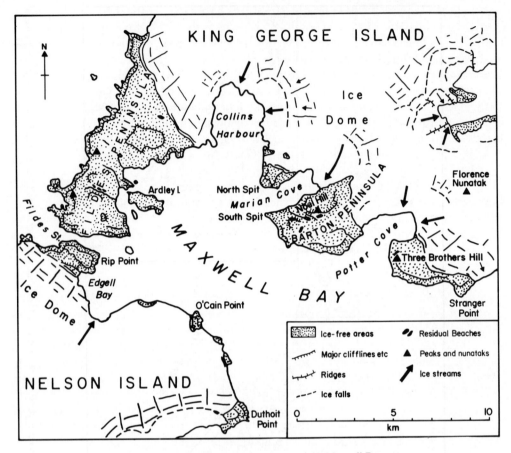

FIGURE 2. Generalized morphological map of the Maxwell Bay area

flanked to the west by Nelson Island and to the north and east by King George Island, and it is connected with the waters of Drake Strait through the precipitous gash of Fildes Strait. The bay allows access to a narrow belt of ice-free land ranging from the Rip Point peninsula in the west to Stranger Point in the east (Fig. 2). This belt is broken only by Fildes Strait and by the outlet-glacier troughs of Collins Harbour, Marian Cove and Potter Cove. There is abundant evidence of glaciation on the ice-free peninsulas, and the flanks of the outlet-glacier troughs on the northern side of the bay provide good evidence for the recent retreat of glacier fronts and for the associated process of isostatic recovery. Lateral moraines and raised beach-ridges are abundant.

The area under consideration has received little previous attention from geomorphologists. The work of J. E. Ferrar (1962), H. A. Orlando (1964), O. C. Schauer and N. H. Fourcade (1964) and Barton (1965) is largely concerned with solid geology, and Maxwell Bay is mentioned but briefly in the review articles of Adie (1964a; 1964b). Only recently has the geomorphology of the central South Shetland Islands been discussed in more detail (John and Sugden, 1971; K. R. Everett, 1971).

MAXWELL BAY AND ITS SUBMARINE TROUGH

At the present day the form of Maxwell Bay is deceptively simple, for it appears to be but a shallow embayment flanked by an undulating or gently stepped coastline which is partially ice-covered. The presence of fluvially-shaped erosion surfaces and ancient marine plat-forms both above and below sea level indicates that the Bay probably originated in pre-Pleistocene times. Apart from Fildes Strait, the deeper coastal embayments can be related to the outlets of ice streams draining from the King George Island and Nelson Island ice caps. The greater part of the coastal zone lies below an altitude of 150 m, and indeed the highest point around Maxwell Bay is Noel Hill on Barton Peninsula, with an altitude of only 295 m. On the other hand, Admiralty Chart No. 1774 (1962) shows a considerable trough which lies completely below sea level. This feature is cut into an undulating sub-merged erosion surface which sometimes extends to a depth of 128 m. To the east of Ardley Island there is a steep drop at the trough head from 177 m to 408 m, and a depth of 484 m is recorded about 2 km to the south. Off Duthoit Point there is a steep fall at the trough edge from 44 m to 539 m, over a distance of less than 3 km. For some distance beyond the mouth of Maxwell Bay the trough is deeply incised into the submerged offshore shelf, attaining a depth of more than 730 m at its exit on the north slope of Bransfield Strait (Fig. 3).

In view of the apparently straight sides of the Maxwell Bay trough it is worth con-sidering the possibility that it is a rift valley bounded by parallel faults. This hypothesis is attractive, for it is known from geophysical investigations that there has been vertical dis-placement of some 2000 m on a major fault running along the northern side of Bransfield Strait (D. H. Griffiths *et al.*, 1964; F. J. Davey and R. G. B. Renner, 1969). As recognized by O. Nordenskjöld as long ago as 1913, the projected line of this fault coincides with the recent volcanic vents at Deception Island, Penguin Island and Bridgeman Island. However, both this fault and other faults recognized by G. J. Hobbs (1968) on Livingston Island and by D. D. Hawkes (1961) and Barton (1965) on King George Island are strike faults aligned approximately parallel with the south-west to north-east trend of the central South Shet-land Islands themselves. There is no evidence of cross-faulting on a scale sufficient for the formation of the Maxwell Bay trough; those faults which do run across the axis of Fildes Peninsula are associated with small-scale downthrows (Schauer and Fourcade, 1964; Barton, 1965), and there is disagreement about whether the central block of the peninsula is up-faulted or down-faulted.

The submarine trough of Maxwell Bay is but one of seven such features cut into the south-eastern flank of the central South Shetland Islands. Of these, only Admiralty Bay has the appearance of a true fjord at the present day (Linton, 1964). The others are in the major straits separating the islands, in King George Bay, and south-west of False Bay. At least four of the troughs are over-deepened. Without recognizing the presence of the submerged troughs, Holtedahl (1929) concluded 'There can be no doubt that the sounds separating the South Shetland islands are formed . . . by glacial transverse erosion and we must therefore assume that these depressions have once been ice filled' (p. 94). It seems most likely that, like the others, the Maxwell Bay trough is a glacially eroded feature, cut by a powerful ice stream flowing south-eastwards along a pre-existing embayment. Somewhat surprisingly, the Marian Cove and Potter Cove outlet glaciers appear to have had little influence upon the form of the trough, for both coves are relatively shallow features.

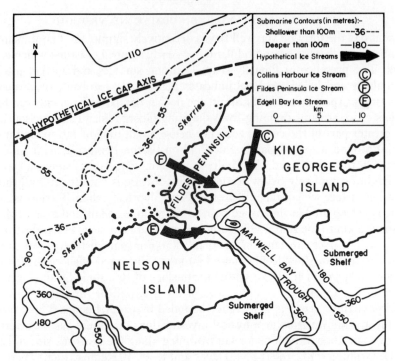

FIGURE 3. The submerged trough in Maxwell Bay and hypothetical directions of ice movement during the stage of maximum glaciation. (Submarine contours are interpolated from Admiralty Chart No. 1774)

THE EARLIEST GLACIAL STAGE

A major problem concerns the source of the ice which cut the Maxwell Bay trough. It cannot have been derived entirely from the narrow arc of land around the trough head, for neither Fildes Peninsula nor the Rip Point peninsula are extensive enough or high enough to have supported an active ice dome. It is suggested, therefore, that the ice cap which supplied the Maxwell Bay glacier must have been grounded to the *north-west* of the islands, on the skerries and on the wide submarine shelf farther north (Fig. 3). This suggestion is supported by the evidence from the other troughs of the island group (John and Sugden, 1971), and by the following morphological characteristics of Fildes Peninsula and the Rip Point peninsula:

(1) Both areas appear to be ice-moulded. Remnants of erosion surfaces have irregular hummocky rock surfaces and stand as convex hillocks above elongated hollows often occupied by lakes. Smoothed bedrock slabs are frequently encountered, and here and there narrow gullies cross hills and hollows indiscriminately. While the streamlining effect may have been reinforced by the structure of the area (Schauer and Fourcade, 1964; Barton, 1965), the landscape as a whole is reminiscent of areas of 'knock and lochan' relief in Scotland (Fig. 4). On Fildes Peninsula this type of relief is strikingly well-developed in the broad central depression which runs north-west to south-east across the waist of the peninsula, separating the northern and southern upland blocks. The overall trend of the streamlined features is west-north-west to east-south-east, a direction which is consistent with the idea of ice overriding the Fildes Peninsula–Rip Point area before entering the Maxwell Bay trough.

FIGURE 4. 'Knock and lochan' relief in the central depression of Fildes Peninsula, King George Island. View towards the south. In the distance can be seen the smooth ice dome of Nelson Island. (*Reproduced by the courtesy of the British Antarctic Survey*)

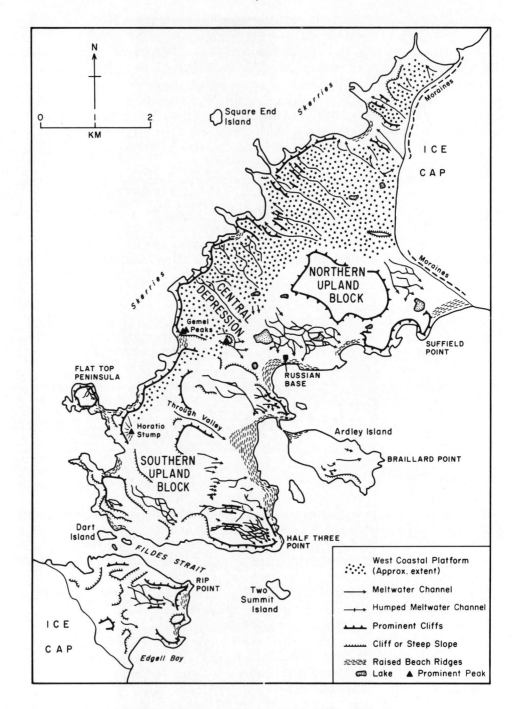

FIGURE 5. Generalized morphological map of Fildes Peninsula

(2) Meltwater channels appear to be uncommon in Antarctica and even on the sub-Antarctic islands, and it was surprising to discover several superb channel systems at the head of Maxwell Bay (Fig. 5). In the central depression of Fildes Peninsula, several channels have their intakes on the western side of the watershed and cross cols through humped sections at 40–55 m. They then split up into complex anastomosing systems in which individual channels are up to 15 m deep, with flat floors and occasional cliffed sides. One such system is located just to the north of Bellingshausen Base. As with the channel systems on the southern upland block, the conclusion is inescapable that the erosive agent was subglacial meltwater, flowing generally south-eastwards under the influence of an ice gradient falling across the watershed in that direction.

Thus the evidence of both ice moulding and meltwater channel erosion at the head of Maxwell Bay seems to indicate that, although the Bay may have been influenced originally by tectonic, fluvial and marine processes, it owes its present form largely to the erosive action of ice. The glacier responsible originated in a source area that is no longer above sea-level, and moved south-eastwards across Fildes Peninsula and Rip Point. The central depression of Fildes Peninsula has been scoured by this major ice stream *en route* for the trough head. The coastal embayments of Collins Harbour and Edgell Bay may have originated as a result of excavation by tributary ice streams at the maximum stage of glaciation, for they contain submerged bifurcations of the Maxwell Bay trough head; but Marian Cove and Potter Cove are relatively shallow features which may be of more recent date.

THE NON-GLACIAL INTERVAL IN MAXWELL BAY

The maximum stage of glaciation in Maxwell Bay was followed after an unknown time interval by a phase of raised-beach formation. Pockets of well-rounded igneous pebbles are found in sheltered localities inland on Fildes Peninsula and Barton Peninsula and near Three Brothers Hill (Fig. 2). In some places the deposits are banked against small clifflines cut into glacially eroded rock knolls, and they seem to rest upon narrow rock benches. Their distribution is quite unrelated to the presence of meltwater channels, and there is nothing to suggest that they are fluvio-glacial in origin (John and Sugden, 1971). They are strikingly similar to some of the present-day beaches on the island coasts. They are excellently preserved except where disturbed by solifluction.

The beaches have undoubtedly been affected by overriding ice since they were formed, for there are erratic 'fans' of striated cobbles which in some cases can be traced back to the beach source in a sheltered locality. For this reason the beaches are referred to as *residual beaches*, in the conviction that the few remaining *in situ* deposits are but the remnants of a formerly more extensive set of beach ridges.

By far the greatest problem of interpretation is concerned with the altitudes of the residual beaches. Of the ten beach localities discovered, there are two at an altitude of 55 m and a further four between 100 m and 160 m. The highest occurs at an altitude of 275 m on the flank of Noel Hill. So far as the author is aware, these are the highest yet discovered in Antarctica. In addition, they may be the only beaches on the continental periphery which can be assigned with any confidence to a Pleistocene non-glacial interval which was followed by local ice expansion (Calkin and Nichols, 1970).

The beach altitudes can be explained by the following hypotheses, singly or in combination:

(a) tectonic uplift since the beaches were formed;

(b) early Pleistocene sea-levels considerably higher than the present;

(c) large-scale isostatic recovery of the South Shetland Islands following the end of the maximum glaciation;

(d) isostatic uplift on the continental periphery in compensation for a glacially-induced depression of the continental interior.

Hypothesis (a) is not favoured, for the residual beaches are relatively fresh features too young to have been affected by the last major phase of tectonic activity (Adie, 1964b). Hypothesis (b) appears unreliable in view of the lack of evidence for any Pleistocene interglacial sea-level as high as 275 m (A. Guilcher, 1969). Hypothesis (c) has more to commend it. However, isostatic recovery to an altitude of 275 m must have involved original crustal depression of about 400 m, taking into account glacio-eustatic sea-level lowering. The South Shetland Islands ice-cap can hardly have been large enough to achieve this (W. S. Broecker, 1966). On balance, it seems most likely that the high residual beaches of the Maxwell Bay area owe their considerable altitude to compensational uplift or 'marginal bulge', triggered off by an increase in ice volume in the main Antarctic ice sheet and by depression of the interior parts of the continent (P. S. Voronov and S. A. Ushakov, 1965; Hollin, 1970b). Possibly, therefore, although the beaches may be interglacial or interstadial features, their uplift may be related to a succeeding glacial stage.

THE LAST EXPANSION OF THE ISLAND ICE CAPS

As noted above, there has undoubtedly been a stage of ice expansion since the formation of the residual beaches. Most of the ice-free peninsulas have a scatter of erratics and a patchy cover of till containing faceted, grooved and striated stones of all shapes and sizes. On Duthoit Point there is a veneer of loose earthy till well beyond the present ice edge on the undulating surface at $c.$ 78 m, and on Rip Point and O'Cain Point there is frost-heaved till on platform remnants at $c.$ 40 m. In the latter case it is significant that the till contains abundant well-rounded and striated cobbles which have apparently been derived from pre-existing beach deposits. Similarly, Fildes Peninsula and Barton Peninsula have a sporadic till cover containing a few well-rounded cobbles; there is also a scatter of erratics, and perched blocks are frequently encountered. On slopes the till has been redistributed by solifluction processes, and on the flatter surfaces it has been subjected to cryoturbation. While the recurrence of rock outcrops across Fildes Peninsula makes the tracing of erratic sources somewhat difficult (Barton, 1965), it is noticeable that the distribution of the till cover is indiscriminate; till is found as a veneer over the whole of the ice-scoured area in the central depression, and also within meltwater channels wherever they occur. Those localities devoid of till are mainly coastal embayments where marine processes have been active since the retreat of the ice; in these embayments raised beaches are common at $c.$ 18·5 m and below (Fig. 5). On some peninsulas remote from the ice margin there is a washing-limit at $c.$ 54 m.

It is unfortunate that on Fildes Peninsula there are no terminal morainic ridges or other retreat features that might enable one to infer the source of the ice of the last glaciation or its direction of retreat. However, two lines of evidence are useful in this context:

(1) Over most of the extensive 35–48 m platform on the western side of the peninsula, and in the deep coastal channels, there is a continuous fresh till sheet full of striated and faceted erratic pebbles. The till is compact and sticky with a silt and clay matrix, indicating that it has not been washed or otherwise modified by marine processes. The same may be

FIGURE 6. The washed outer edge of the 35–48 m coastal platform on the western side of Fildes Peninsula. Photograph taken near Gemel Peaks. Note the absence of till and the scatter of erratic stones. (*Reproduced by the courtesy of the British Antarctic Survey*)

said of Square End Island (Fig. 5), whose upper surface (above 33 m) has a veneer of fresh till. In contrast, the platform to the south-east of Gemel Peaks has a washed appearance (Fig. 6). Till patches are rarely encountered, and pockets of washed till are concentrated along gullies and channels cut into the platform surface. This contrast cannot be explained by any difference of altitude between the northern and southern sections of the platform, for it is essentially horizontal over a distance of more than 8 km. It seems, therefore, that at some stage following the peak of the last glaciation, an ice front (which was approximately parallel with the present edge of the King George Island ice cap) crossed the west coast of Fildes Peninsula in the vicinity of Gemel Peaks while relative sea level lay at an altitude of *c*. 35 m. By the time that the ice front had retreated from this position, sea level had fallen to such an extent that it was not able to affect the northern part of the platform. It is significant that the only raised beach remnant above 18·5 m on the west coast of Fildes Peninsula occurs at *c*. 38 m near Horatio Stump, *c*. 2 km south of Gemel Peaks (John and Sugden, 1971). This evidence strongly suggests that the last glaciation of Fildes Peninsula was entirely the result of the local expansion of the King George Island ice cap.

(2) From the mapping of the residual beaches referred to above, it was discovered that most of the remnants are situated beneath west- or south-west-facing cliff notches; this is precisely the orientation which would afford the greatest protection from ice moving *along* the axes of the ice-free peninsulas. This is confirmed by the orientations of the erratic fans emanating from the residual beaches. On the other hand, the inability of the ice of the last glaciation to remove all traces of the residual beaches or to achieve any erosional modification of the streamlined features orientated *across* peninsula axes, suggests a relatively feeble and possibly short-lived local glaciation. In other words, this glaciation was distinct both in magnitude and direction of ice movement from the maximum glaciation.

THE RECENT RETREAT OF THE ISLAND ICE CAPS

Where the present-day ice-cap margins terminate on land they are often marked by pro-

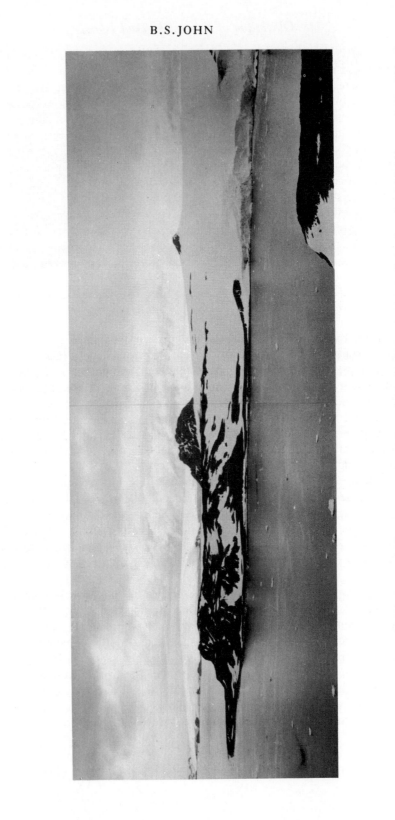

FIGURE 7. View towards the north across the Marian Cove outlet glacier trough. On the ice-free peninsula two distinct lateral morainic ridges can be seen close to present sea level. On the extreme left of the photograph can be seen the edge of the King George Island ice cap on Fildes Peninsula. (*Reproduced by the courtesy of the British Antarctic Survey*)

minent morainic ridges. Examples may be seen on Fildes Peninsula, where there is a series of discontinuous morainic ridges at the ice edge, flanked by a snow ramp. There are no ice-edge moraines on Barton Peninsula or Duthoit Point, but there are three prominent morainic ridges on O'Cain Point, and a broad zone of spectacular ice-cored moraine to the east of Three Brothers Hill. The crests of these morainic ridges stand up to 10 m above the adjacent ice surfaces, indicating a recent slight lowering of the ice-cap margins.

The most convincing evidence of recent ice-margin fluctuations is to be seen in the outlet-glacier troughs of Marian Cove and Potter Cove. Here lateral moraines and raised beaches are intimately related (Fig. 7), and it is suggested that, by about 10 000 years BP, the outlet glaciers were well withdrawn into their troughs. Raised beaches were formed as relative sea level fell from *c.* 54 m. Continued ice-edge retreat and isostatic recovery led to successive phases of beach formation, with an important stillstand or marine transgression represented at *c.* 18·5 m. On the basis of radiocarbon age determinations from Potter Cove and elsewhere, it has been suggested that a later local glacier readvance culminated about 500 years BP at the mouths of the bays. This readvance may be the equivalent of the 'False Bay glacial event' recognized by Everett (1971) from Livingston Island. Subsequently the glacier fronts have retreated gradually, accompanied by isostatic uplift and the formation of a succession of raised beaches from altitudes of *c.* 6 m at the bay mouths to *c.* 3 m close to the present glacier snouts.

It is difficult to establish links between the lateral moraine ridges of the outlet glacier troughs and the ice-margin moraines referred to earlier. However, the zone of ice-cored moraine east of Three Brothers Hill (Fig. 2) runs directly into one of these ridges, and it appears likely that all of these features are related to the same short-lived climatic deterioration.

DISCUSSION

The foregoing data assembled from the Maxwell Bay area suggest the glacial sequence outlined on Table I. This sequence is based for the most part upon morphological evidence, and its reliability (in terms of an absolute time-scale) cannot therefore compare with the sequence being worked out for the McMurdo region (Denton *et al.*, 1969). Nevertheless, the upper part of the sequence now appears to be adequately fixed by radiocarbon age determinations, and the lower part is quite compatible with other evidence from the Antarctic Peninsula. In particular, the whole sequence deserves further consideration in the light of Mercer's hypothesis mentioned above.

The maximum glaciation (or glaciations) (1A) appears to have been responsible for the bulk of glacial erosion in the Maxwell Bay area and elsewhere in the central South Shetland Islands. At this time an ice cap accumulated with dimensions approximately 250 km × 65 km and based for the most part on the shelf now submerged to the north-west of the island group. This ice cap may have developed from the grounding of a small ice shelf, which would imply that the mean annual air temperature for the warmest month was less than 0°C, and that the ice would have been polar rather than sub-polar or temperate in character (Robin and Adie, 1964). Alternatively, the ice cap may have grown by the coalescence and northward expansion of island ice-cap peripheries in response to a glacio-eustatic lowering of sea level. A sea level even 110 m lower than the present would have exposed the greater part of the wide submarine shelf. As the ice cap became firmly established on this shelf its axis moved gradually northwards towards the source of the precipitation that nourished it. There is no evidence that the South Shetland Islands were ever overwhelmed by ice moving north from the tip of the Antarctic Peninsula, for the depths of Bransfield Strait would have presented an insurmountable obstacle. On the other hand, the islands' ice cap may have been linked to that of the Antarctic Peninsula across the submerged shelf running from Anvers Island to Low Island and Smith Island (Voronov, 1965).

TABLE I

A suggested glacial sequence for the Maxwell Bay area

Stages and sub-stages	Associated events	North European correlatives
(4B) Glacier retreat	Raised beaches below 6 m	500 yrs BP–present*
(4A) Readvance of outlet glaciers	Morainic ridges in troughs	800–500 yrs BP*
(3B) Deglaciation	(d) Isostatic recovery-raised beaches below 54 m	} Flandrian (post 10 000 yrs BP)*
(3A) Local glaciation	(c) Veneer of till, erratics, etc. (b) Destruction of many raised beaches, solifluction deposits, etc. (a) Expansion of island ice caps over Fildes Peninsula and other peninsulas	} Weichselian
(2) Non-glacial interval	Residual beaches formed up to 275 m Some marine erosion of pre-existing glacially-scoured relief	} Eemian
(1B) Deglaciation	Rapid ice dissolution and cutting of meltwater channels	
(1A) Maximum glaciation (or glaciations)	(c) Glacial erosion and streamlining of bedrock (b) Cutting of glacial troughs (a) Inundation of Bay peripheries by ice moving from north-western ice cap	} Saalian or earlier

* Radiocarbon ages

Once the ice cap was fully developed, radiating ice streams were responsible for the cutting of the Maxwell Bay and other troughs and for the scouring of bedrock relief on Fildes Peninsula and elsewhere. The magnitude of this erosion may indicate that the glaciation spanned a great period of time, and that the ice streams were occasionally wet-based and thus temperate or sub-polar in type. This would be predictable from the peripheral and oceanic location of the South Shetlands; it need not necessarily be assumed that the maximum phase of erosion in the islands was contemporaneous with the early phase of alpine erosion by wet-based glaciers in other parts of West Antarctica (Mercer, 1968c).

While the foregoing account of the 'maximum glaciation' may be an over-simplification of a whole sequence of glacial stages spanning many thousands of years, the evidence does not at present permit any reliable subdivision to be made. However, there is ample evidence for the rapid dissolution of the South Shetland Islands ice cap (1B and 2). This may have been triggered off by an interglacial eustatic rise of sea level. On the other hand, it appears more likely that climatic warming was the cause of ice dissolution, for the meltwater channels of Fildes Peninsula must have been cut by large volumes of sub-glacial meltwater draining towards the head of the Maxwell Bay trough. As suggested on page 84 one interpretation of the residual beaches of Maxwell Bay would also involve rapid (and possibly complete) dissolution of the ice cover. It is tempting to relate this phase to the catastrophic dissolution of the West Antarctic ice sheet, referred by Mercer (1968a) to the last interglacial. From the South Shetlands evidence, an Eemian age appears entirely reasonable on the following grounds:

(1) If the meltwater channels and residual beaches are any older, they would have been subjected to much greater denudation during at least two stages of glaciation.

(2) If the features are any younger, they would have to be assigned either to the mid-Weichselian interstadial or to the Flandrian. In the former case it would be difficult to envisage climatic conditions warm enough for the rapid dissolution of the South Shetland Islands ice cap (Epstein *et al.*, 1970) and in the latter case it would be difficult to envisage a later climatic deterioration on a scale large enough to inundate previously ice-free peninsulas with local ice.

The rapid wastage of the South Shetland Islands ice cap during the last interglacial was followed by a renewed expansion of ice from smaller ice caps centred, as now, on the individual islands of the group (3A). On Livingston Island this glaciation is referred to by Everett (1971) as the 'Livingston glacial event'. It is not yet possible to determine the outermost limits of this Weichselian glaciation, but it seems to have achieved relatively little in the way of landscape modification. Possibly the island ice caps consisted of polar ice or, more likely, they experienced much lower annual precipitation totals than the north-west ice cap of the Saalian stage; in either case glacier activity and the erosive capacity would have been adversely affected. More obviously, as in the northern hemisphere, the Weichselian glaciation was simply too short-lived to have been fully effective.

As the climatic warming of the Flandrian stage set in there was a gradual retreat of ice-cap margins (3B). Meltwater production was minimal; as a result, the ice-free peninsulas have no fresh (i.e. drift-free) meltwater channels and no recognizable deposits of fluvio-glacial sands and gravels. The only readvance stage for which there is clear evidence (4A) culminated at about 500 years BP, some time earlier than the 'Little Ice Age' or Neoglacial readvances recorded from other parts of the world in recent centuries (J. M. Grove, 1966; S. C. Porter and Denton, 1967; Mercer, 1971).

In conclusion, it seems that the suggested glacial sequence for the Maxwell Bay area may be of some value as a working hypothesis for other parts of West Antarctica. In several respects this sequence provides support for Mercer's proposed scheme of events. However, the story needs to be tested by reference to other studies: for example, the mineralogy of sea-floor sediments around the South Shetland Islands (D. S. Edwards and H. G. Goodell, 1969; D. A. Warnke, 1970) and the foraminiferal biostratigraphy of sub-Antarctic deep-sea cores (J. P. Kennett, 1970). In particular, the applicability of the proposed sequence to regions farther south needs to be tested by geomorphological investigations and by a programme of absolute dating.

ACKNOWLEDGEMENTS

The writer cheerfully acknowledges his debt to Dr David Sugden, who did a considerable amount of the work on which this paper is based. Thanks are also due to the British Antarctic Survey, who financed the fieldwork; to Dr R. J. Adie for his constant interest; and to Professor F. W. Shotton of Birmingham University for the radiocarbon dating of the Maxwell Bay organic samples. Dr John Mercer and my colleague Ian Evans have made many invaluable comments upon the paper. The University of Durham Research Fund kindly contributed towards the cost of illustrations.

REFERENCES

ADIE, R. J. (1964a) 'Geological history' in *Antarctic research* (ed. R. E. PRIESTLEY, R. J. ADIE and G. DE Q. ROBIN) 118–62

ADIE, R. J. (1964b) 'Sea-level changes in the Scotia Arc and Graham Land' in *Antarctic geology* (ed. R. J. ADIE), 27–32

BAKER, P. E., T. G. DAVIES and M. J. ROOBAL (1969) 'Volcanic activity at Deception Island in 1967 and 1969', *Nature, Lond.* 224 (5219), 553–60

BARTON, C. M. (1965) 'The geology of the South Shetland Islands: III. The stratigraphy of King George Island', *Br. Antarct. Surv. scient. Rep.* 44, 33 pp.

BROECKER, W. S. (1966) 'Glacial rebound and the deformation of proglacial lakes', *J. geophys. Res.* 71, 4777–83

CALKIN, P. E. and R. L. NICHOLS (*in press*) 'Quaternary studies in Antarctica' in *Antarctic geology and geophysics* (ed. R. J. ADIE), Oslo, Universitetsforlaget

DAVEY, F. J. and R. G. B. RENNER (1969) 'Bouger anomaly map of Graham Land', *Br. Antarct. Surv. Bull.* 22, 77–81

DENTON, G. H., R. L. ARMSTRONG and M. STUIVER (1969) 'Histoire glaciaire et chronologie de la région du détroit McMurdo, sud de la Terre Victoria, Antarctide: note préliminaire', *Rev. Géogr. phys. Géol. dyn.* 11, 265–78

EDWARDS, D. S. and H. G. GOODELL (1969) 'The detrital mineralogy of ocean floor surface sediments adjacent to the Antarctic Peninsula, Antarctica', *Marine Geol.* 7, 207–34

EPSTEIN, S., R. P. SHARP and A. J. GOW (1970) 'Antarctic ice sheet: stable isotope analysis of Byrd Station cores and interhemispheric climatic implications', *Science, N.Y.* 168, 1570–2

EVERETT, K. R. (1971) 'Observations on the glacial history of Livingston Island', *Arctic* 24, 41–50

FERRAR, J. E. (1962) 'The raised beaches of the Noel Hill peninsula, King George Island, South Shetland Islands', *Br. Antarct. Surv. prelim. geol. Rep.* 13, 12 pp. (unpubl.)

FLEMING, W. L. S. (1940) 'Relic glacial forms on the western seaboard of Graham Land', *Geogrl J.* 96, 93–100

GRIFFITHS, D. H., R. P. RIDDIHOUGH, H. A. D. CAMERON and P. KENNETT (1964) 'Geophysical investigation of the Scotia Arc', *Br. Antarct. Surv. scient. Rep.* 46, 43 pp.

GROVE, J. M. (1966) 'The Little Ice Age in the Massif of Mont Blanc', *Trans. Inst. Br. Geogr.* 40, 129–43

GUILCHER, A. (1969) 'Pleistocene and Holocene sea level changes', *Earth Sci. Rev.* 5, 69–97

HAWKES, D. D. (1961) 'The geology of the South Shetland Islands: I. The petrology of King George Island', *Scient. Rep. Falkd Isl. Depend. Surv.* 26, 28 pp.

HOBBS, G. J. (1968) 'The geology of the South Shetland Islands: IV. The geology of Livingston Island', *Br. Antarct. Surv. scient. Rep.* 47, 34 pp.

HOLLIN, J. T. (1969) 'Ice surges and the geological record', *Can. J. Earth Sci.* 6 (4), 903–10

HOLLIN, J. T. (1970a) 'Antarctic ice surges', *Antarctic J. U.S.* 5 (5), 155–6

HOLLIN, J. T. (1970b) 'Is the Antarctic ice sheet growing thicker?' in *International Symposium on Antarctic Glaciological Exploration (ISAGE)* (ed. A. J. Gow et al.), I.A.S.H. Publ. No. 86, 363–74

HOLTEDAHL, O. (1929) 'On the geology and physiography of some Antarctic and sub-Antarctic islands', *Scient. Results Norw. Antarct. Exped.* 3, 172 pp.

JOHN, B. S. and D. E. SUGDEN (1971) 'Raised marine features and phases of glaciation in the South Shetland Islands', *Br. Antarct. Surv. Bull.* 24, 45–111

KENNETT, J. P. (1970) 'Pleistocene paleoclimates and foraminiferal biostratigraphy in sub-antarctic deep-sea cores', *Deep Sea Res.* 17, 125–140

LINTON, D. L. (1964) 'Landscape evolution' in *Antarctic research* (ed. R. E. PRIESTLEY, R. J. ADIE and G. DE Q. ROBIN), 85–99

MERCER, J. H. (1968a) 'Antarctic ice and Sangamon sea-level', *Bull. int. Ass. scient. Hydrol.* 79, 217–25

MERCER, J. H. (1968b) 'Glacial geology of the Reedy Glacier area, Antarctica', *Bull. geol. Soc. Am.* 79, 471–86

MERCER, J. H. (1968c) 'The discontinuous glacio-eustatic fall in Tertiary sea level', *Palaeogeogr. Palaeoclimatol. Palaeoecol.* 5, 77–85

MERCER, J. H. (1971) 'Variations of some Patagonian glaciers since the Late-glacial: II', *Am. J. Sci.* 269, 1–25

NICHOLS, R. L. (1947) 'Elevated beaches of Marguerite Bay, Antarctica', *Bull. geol. Soc. Am.* 58, 1213

NORDENSKJÖLD, O. (1913) 'Antarktis' in *Handbuch der Regionalen Geologie* (ed. G. STEINMANN and O. WILCKENS), Bd. 8, Abt. 6, Heft 15, 29 pp.

ORLANDO, H. A. (1964) 'The fossil flora of the surroundings of Ardley Peninsula, 25 de Mayo Island, South Shetland Islands' in *Antarctic geology* (ed. R. J. ADIE), 629–36

PORTER, S. C. and G. H. DENTON (1967) 'Chronology of Neoglaciation in the North American cordillera', *Am. J. Sci.* 265, 177–210

ROBIN, G. DE Q. and R. J. ADIE (1964) 'The ice cover', in *Antarctic research* (ed. R. E. PRIESTLEY, R. J. ADIE and G. DE Q. ROBIN), 100–117

RUTFORD, R. H., C. CRADDOCK and T. W. BASTIEN (1968) 'Late Tertiary glaciation and sea-level changes in Antarctica', *Palaeogeogr. Palaeoclimatol. Palaeoecol.* 5, 15–39

SCHAUER, O. C. and N. H. FOURCADE (1964) 'Geological-petrographical study of the western end of 25 de Mayo Island, South Shetland Islands' in *Antarctic geology* (ed. R. J. ADIE), 487–91

VORONOV, P. S. (1965) 'Attempt at a reconstruction of the Antarctic ice cap of the epoch of the Earth's maximum glaciation', *Inf. Byull. sov. antarkt. Eksped.* III, 88–93

VORONOV, P. S. and S. A. USHAKOV (1965) 'Some problems in studying isostatic processes in Antarctica', *Inf. Byull sov. antarkt. Eksped.* III, 351–4

WARNKE, D. A. (1970) 'Glacial erosion, ice rafting, and glacial-marine sediments: Antarctica and the Southern Ocean', *Am. J. Sci.* 269, 276–94

RÉSUMÉ. *Au sujet d'une histoire glaciaire de l'Antarctique de l'Ouest.* L'histoire glaciaire de la région de McMurdo est maintenant assez bien connue, mais on n'a trouvé que peu de renseignements sûrs sur la chronologie glaciaire de l'Antarctique de l'Ouest. D'autre part J. H. Mercer a avancé qu'une calotte glaciaire de l'époque Saalienne s'effondra d'une façon désastreux pendant le Riss-Würm inter-glaciaire et fut suivie d'une nouvelle calotte glaciaire pendant la période Weichsélienne. On a donné des preuves dans la région de la Baie de Maxwell des Îles Shetland du Sud par soutenir cette hypothèse. Pendant une période de glaciation très ancienne (Saalienne?) l'extrémité de la baie de Maxwell fut submergeé par des glaces qui, provenant d'une calotte appuyée sur la plate-forme sous-marine peu profonde, se dirigeaient vers le sud-est. Ces glaces furent la cause du profil de la topographie de la couche de pierre sur la péninsule de Fildes et de l'érosion de la vallée glaciaire qui est maintenant submergée dans la Baie de Maxwell. La calotte s'amincit rapidement quand une période plus chaude arriva, ce qui entraîna l'érosion des systèmes de canaux anastomosants par des glaces tondues venant d'au-dessous du glacier. L'altitude des rivages exhaussés autour de la Baie de Maxwell est probablement dûe à un soulèvement plus tardif et compensateur qui accompagna la nouvelle calotte glaciaire et la dépression isostatique à l'intérieur du continent. Pendant la période Weichsélienne il y eut une glaciation locale et d'intensité limitées pendant laquelle des glaces des calottes des Îles recouvrirent des péninsules, qui jusque-la avaient été exemples de glace, et enlevèrent la plupart des rivages exhaussés qui existaient auparavant. La régression du front du glacier pendant l'holocène et le relèvement isostatique furent accompagnés par la formation de rivages exhaussés au-dessous de 50 m et il y eut une nouvelle avancée du glacier de courte durée qui atteignit son maximum à environ il y a 500 ans. Cette suite d'événements pourrait être représentatif d'une bien plus caste région et en particulier du secteur de la Péninsule antarctique de l'Atlantique de l'Ouest.

FIG. 1. Les Îles Shetland du Sud et leur situation en ce qui concerne l'extrémité de la Péninsule Antarctique. Le carré marqué autour de la Baie de Maxwell indique la région que couvre la figure 2

FIG. 2. Carte morphologique générale de la région de la Baie de Maxwell

FIG. 3. La vallée submergée dans la baie de Maxwell et la direction supposée du mouvement des glaces pendant l'ère de la plus grande glaciation. (Les contours sous-marins sont insérés à partir de l'Admiralty Chart No. 1774)

FIG. 4. La topographie de « Knock et Lochan » dans la dépression centrale de la péninsule de Fildes, l'Île du Roi Georges—vue vers le Sud. Au loin on peut voir le dôme arrondi et glace de l'Île de Nelson

FIG. 5. Carte morphologique générale de la Péninsule de Fildes

FIG. 6. Le contour extérieur balayé par la mer de la plate-forme continentale haute de 35–84m sur la face à l'ouest de la Péninsule de Fildes. Photographe prise près des Sommets de Gemel. Remarquez l'absence de morain de fond et l'éparpillement de blocs erratiques

FIG. 7. Débouché de la vallée du glacier de l'Anse de Marian : vue vers le nord

ZUSAMMENFASSUNG. *Zeugnisse aus den Süd-Shetlandinseln als Beitrag zu einer Geschichte der Vergletscherung in West-Antarktika.* Die Geschichte der Glazialperiode der McMurdogegend ist jetzt verhältnismässig bekannt; aber zu eine Chronologie der Vergletscherung von West-Antarktika lag bis jetzt nur wenig zuverlässiges Material vor. Doch hat J.H. Mercer vermutet, dass ein aus der Saale-Eiszeit stammendes Eisfeld in West-Antarktika während der Eem-Interglazial katastrophal zusammenstürzte, und dass die Eisfläche sich während der Weichsel-Eiszeit wieder bildete. Die jetzige Arbeit legt Zeugnisse aus der Gegend um Maxwell Bay (Mittel-Süd-Shetlandinseln) vor, als Beitrag zur Bestätigung dieser Hypothese. Im Laufe einer frühen (Saale?) Glazialperiode wurde das Vorgebirge von Maxwell Bay von Eismassen überströmt die sich in südöstliche Richtung von einer Eiskappe bewegten, die auf der flachen untermeerischen Bank nördlich der Süd-Shetlandinseln gegründet war. Diese Eismassen waren es, die die Formen des Grundgebirges auf dem Fildes-Halbinsel nach ihren Stromlinien gestalteten, und die das jetzt in Maxwell Bay ertrunkene Glazialbecken erodierten. Die Eiskappe schmolz sehr schnell in Beginn einer wärmeren (Eem?) Periode, was zur Erosion der anastomosierenden Kanalsysteme durch subglaziale Schmelzbäche führte. Die Höhen der Strandreste' (bis 275m), die um Maxwell Bay zu finden sind, erklären sich vielleicht aus der Erhebung, die der Wiederbildung des Eisfelds und der isostatischen Vertiefung im Innern des Kontinents entsprach. Während der Weichselischen Glazialperiode war die Eisbildung nur örtlich und von beschränkter Dauer und Stärke; dabei wurden die bis zu jener Zeit eisfreien Halbinseln von Eismassen von den Inseleiskappen überlaufen, die den grössten Teil der schon existierenden Strandterrassen abhobelten. Isostatische Ausgleichung und Rückgang des Eisrands in der Flandrianischen Zeit haben bis in eine Höhe von 54m die Bildung von gehobenen Strandterrassen verursacht;

ein kurzdauerndes Wiedervorstossen des Eises war um 500 Jahre v.H. wieder zu Ende. Diese Reihenfolge der Geschehnisse dürfte für ein weit grösseres Gebeit typisch sein, insbesonderes für den durch den antarktischen Halbinsel gebildeten Sektor von West-Antarktika.

ABB. 1. Die Süd-Shetlandinseln in ihrem Verhältnis zur Spitze des antarktischen Halbinsels. Eingerahmt: das in Abb. 2 dargestellte Gebiet um Maxwell Bay

ABB. 2. Allgemeine morphologische Karte der Maxwell Bay-Gegend

ABB. 3. Das ertrunkene Glazialbecken in Maxwell Bay, und die vermutlichen Richtungen der Eisbewegung während der Periode maximaler Vereisung. (Untermeerische Konturlinien von Admiraltykarte Nr. 1774)

ABB. 4. ‚Knock and Lochan‘ Topographie der zentralen Vertiefung in dem Fildes-Halbinsel auf dem Insel König Georg. Blick in südliche Richtung; in der Ferne sieht man die glatte Eiskuppel der Insel Nelson

ABB. 5. Allgemeine morphologische Karte des Fildes-Halbinsel

ABB. 6. Der gewaschene Aussenrand der 35-48m-langen Küstenterrasse auf der Westseite des Fildes-Halbinsel. Aufnahme in der Nähe von Gemel Peaks (Gemelhöhen). Bemerkenswert die Abwesenheit von Geschiebelehm und das umhergestreute erratische Geschiebe

ABB. 7. Blick gegen Norden über den Gletschertrog des Ausflusses in die Marienbucht (Marian Bay). Auf der eisfreien Halbinsel sieht man zwei ganz deutliche, durch Seitenmoränen gebildete Rücken, in der Nähe von dem jetzigen Meeresspiegel. An der linken Seite der Aufnahme den Rand des Insels König Georg-Eiskappe auf dem Fildes-Halbinsel

Tors, rock weathering and climate in southern Victoria Land, Antarctica

EDWARD DERBYSHIRE

Senior Lecturer in Geography,
University of Keele

Revised MS received 4 November 1971

ABSTRACT. Morphological and weathering characteristics of a group of tors, possessing both angular and rounded joint-blocks, at Sandy Glacier (Wright valley) in the McMurdo 'oasis' of southern Victoria Land are described. Clay minerals (kaolinite family) at and beneath the surface of the rounded corestones indicate chemical weathering of the dolerite. The juxtaposition of rounded and angular tors on the summit of the arête is explained in terms of local variations in microclimate, especially as it affects snow cover. Both types of tor are the product of the prevailing polar desert conditions. It is concluded that such conditions have probably prevailed here throughout the Pleistocene with only brief episodes of slightly more maritime climate.

TAYLOR and Wright valleys form part of the extensive ice-free 'oasis' of southern Victoria Land, Antarctica. These so-called dry valleys are located close to the present mean position of the northern margin of the Ross Ice Shelf and some 1300 km south of the mean summer limit of the Antarctic pack ice. Both Taylor and Wright valleys are occupied in their upper reaches by outlet tongues of the continental ice sheet, Wright valley being cut off from McMurdo Sound by the Wright Lower Glacier tongue of the Wilson Piedmont Glacier (Fig. 1). Cirque and valley glaciers in the intervening ranges hang above these valleys which are essentially straight trenches 1000 m deep running east-north-east in a landscape with a relative relief of over 2000 m.

The whole of the McMurdo Sound region suffers a polar desert climate with attenuated maritime influences, the dry valleys to the west being rather more continental. While winds in all seasons are predominantly from between north-east and south (*U.S. Navy,* 1965), controlled by convergence of Southern Ocean lows on the Ross Sea embayment, westerlies also occur, descending from the ice plateau to be channelled and reinforced as katabatic winds. These westerlies may be almost as prevalent in the dry valleys as the regional easterlies, to judge from a variety of phenomena including ventifacts, sand sheets and abraded seal carcasses. With Ross Sea depressions to the east moving south and filling, neither easterlies nor westerlies produce much precipitation in the ice-free valleys. This dryness is emphasized in summer by relatively low albedos which also serve to raise both rock surface and maximum air temperatures above the freezing point (F. C. Ugolini and C. Bull, 1965). The prevailing arid periglacial climate may have affected the dry valleys for at least several thousand years before the present (R. L. Nichols, 1963, 1966). Geomorphological and geological evidence suggests that recurrent and prolonged arid periods have characterized these valleys both in interglacial episodes (Nichols, 1965) and following the last deglacierization (A. T. Wilson, 1964). On the other hand, it is becoming clear that this aridity has been punctuated by rather more maritime climatic phases. This has been

93

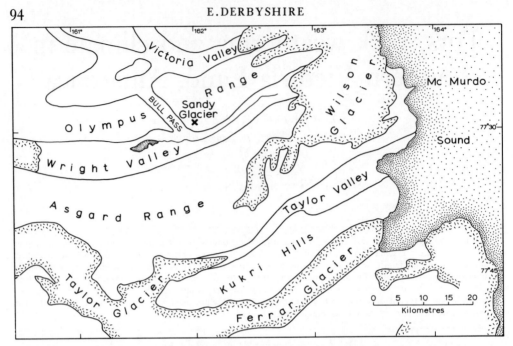

FIGURE 1. Location map of the dry valleys, southern Victoria Land. Sandy Glacier is 100 km west-north-west of McMurdo station, Ross Island

implicit in stratigraphical studies, for example, in the stress laid upon arid periglacial weathering of moraines and erratics as a criterion of age (e.g. C. Bull, B. C. McKelvey and P. N. Webb, 1962), the recognition of former lakes of moderate (E. E. Angino, M. D. Turner and E. J. Zeller, 1962) or great depth (T. L. Péwé, 1960) in Taylor Valley following past glacial advances (but see W. Dort, 1967), and the widespread occurrence of landforms and deposits such as glacial drainage channels, alluvial fans and solifluction sheets demanding more abundant surface water than is produced at the present time (e.g. P. Calkin, 1964). Recently, past advances in the smaller, more sensitive alpine glaciers have been correlated with periods of more oceanic climate on the basis of studies of the structure and stratigraphy of these glaciers (Dort, 1967, 1967a, 1970) and interbedded glacial and volcanic sediments (G. H. Denton, R. C. Armstrong and M. Stuiver 1969). However, little is known about the number, frequency and magnitude of these changes.

The landforms of the dry valleys are made up of two major assemblages: those characteristic of an arid periglacial régime and others which reflect periodically increased surface run-off. The current dominance of salt weathering, frost shatter and thermal contraction as weathering processes and of wind as the most important general agent of transport and erosion is evident in the development of deflation pavements, rippled sand sheets and dunes, local felsenmeers and salt crusts. In this context, it is proposed to examine some morphological and weathering characteristics of tors in the light of the controversy over their possible morphoclimatic significance.

THE TORS OF SANDY GLACIER

At the southern end of Bull Pass (Fig. 1) on the broad arête separating the cirques occupied

FIGURE 2. Uncontrolled air-photo mosaic showing Sandy Glacier and part of the unnamed glacier to the east. The locations of angular (A) and rounded (R) summit tors and hillslope tors in the Beacon Sandstone series (B) are shown, as well as the ice-cored cirque glacier moraines (MM). Orientation approximately north-south. (United States Air Force photographs)

by Sandy Glacier and an unnamed glacier to the east, are found a variety of rock stacks or tors developed in Palaeozoic sandstones (Beacon Sandstone series) and Ferrar dolerites of Jurassic age (Fig. 2). The sandstone tors stand out from the west-facing cliff of the lower slopes of the arête as simple stacks bounded by two sets of nearly vertical joints (Fig. 3). On the same slope above the sandstone, single-pinnacle tors have developed in dolerite. Both the sandstone beds and the third major joint set in the dolerite dip gently towards the west and this has given the arête its marked asymmetry. The sandstone tors and sub-summit dolerite tors are essentially angular in form and are consistent with an origin by slope retreat resulting from frost riving, the slopes below bearing a cover of angular clitter. Both rounded and sharply angular forms occur, however, in both lithologies. The sandstone tors have been rounded by wind action, particularly in the silty members (Fig. 3),

FIGURE 3. Hillslope tors in Beacon Sandstone series on the lower slopes of the Sandy Glacier arête (B in Fig. 2). Arcuate line of dolerite boulders (foreground) is a protalus rampart derived from angular dolerite tors (off photograph to the left). Figure in left-centre middle distance indicates scale. View south-eastward

while the thinly bedded arkoses have been frost-shattered into angular forms. Wind abrasion is particularly noticeable on tor surfaces facing west to south-west. Lower tor slopes and fallen joint blocks of dolerite also show some wind polish and edge-rounding which may be caused by wind action.

The summit tors, developed entirely in the dolerite, are of considerable interest in that they display, within a distance of a few hundred metres, morphological characteristics ranging from extreme angularity (Fig. 4) to the well-rounded joint blocks reminiscent of the corestones associated with tors in more temperate regions. Those summit tors examined are of relatively small size (3–4 m high) and rise from the smooth westerly-sloping ridge crest. This dolerite surface is covered by a veneer of debris made up of frost-shattered joint blocks and a coarse gravel produced by granular disintegration. Ice-wedge polygons cover much of this south-easterly-facing slope as far as the ice-cored moraine of the un-named cirque glacier and are found adjacent to some tors. Farther to the south along the arête, the polygons give way downslope to broad, coarse stone stripes where dolerite gelifractate has moved on to and mingled with the shattered sandstone debris below. The presence of summit tors made up in some cases of angular rock and in most cases of well-rounded joint blocks standing above such a surface of cryoplanation poses a problem in interpretation.

FIGURE 4. Angular, frost-shattered tor in Ferrar dolerite with ice-wedge polygons, ice-cored moraine and unnamed cirque glacier in the background. View northward

The rounded blocks of the summit tors of Sandy Glacier have an iron-stained, flaky surface which can be broken down into gravel and sand-size grains in the hand, although no cavernous weathering or tafoni were noted. Exfoliation shells are common and can be traced down to the shallow permafrost table. Removal of the shells appears to result from frost action which has produced freshly fractured surfaces near the bases of several of the rounded tors. Flaking, exfoliation and granular disintegration, followed by wind removal of the products, have been described from many locations in the dry valleys, particularly in relation to cavernous weathering of granite and gneiss. Such weathering forms have been explained in terms of mechanical processes, particularly the growth of crystals of halite and mirabilite (e.g. A. Cailleux, 1962) which are widespread and locally very abundant. There is some difference of opinion about the role of chemical weathering in southern Victoria Land. While pedologists have established that the production of clay minerals by slow hydration of micas is active (I. B. Campbell and G. G. C. Claridge, 1967) and widespread in Taylor valley (J. D. McCraw, 1967) and Wright valley (K. R. Everett and R. E. Behling, 1970) under the present climate, chemical weathering of rock surfaces has generally been rejected, a view reinforced by a rigorous analysis of a severely weathered quartz diorite at Marble Point by W. C. Kelly and J. H. Zumberge (1961), although work by Glazovskaia (in Russian, quoted by J. C. F. Tedrow and Ugolini (1966)) has detected clays derived from weathering of felspars at Mirny.

The surface condition of the rounded dolerite tors at Sandy Glacier appears similar

FIGURE 5. Rounded ('woolsack') tor in dolerite, showing exfoliation shells (left) and frost-shattered face (right). Samples for mineralogical and weathering studies were taken from this tor

to the later stages in the weathering described in the Marble Point diorite. Lying at essentially the same altitude, the structure of both angular and rounded tors was found to be similar, while their petrography approximates closely to the descriptions given of the lower sill of Wright valley by B. N. Gunn (1962) and McKelvey and Webb (1962), as a medium-grained tholeiitic quartz dolerite with biotite, apatite, chlorite and iron oxides as minor accessory minerals. Thin sections of samples of the flaking margins of corestones showed weathering fractures, generally along crystal boundaries and cleavages, to a depth of at least 5 cm.[1] Cracks locally appear iron-stained (Fig. 6), especially around and within the grains of pyroxene. Fine material is also present in the cracks and on the disintegrating surfaces of the joint blocks; scrapings were taken, roughly ground and subjected to X-ray diffraction. The resulting traces (Fig. 7), in addition to showing clear maxima corresponding with the d-spacings of labradorite, the pyroxenes and quartz, show clear peaks at about 0·72 nm, 1·01 nm and 1·21 nm. The first peak is interpreted as kaolinite in view of the good secondary peak at 0·315 nm, the second peak may correspond with the d-spacing of halloysite, another clay mineral of the kaolinite family, while the 1·21 nm peak is proble-

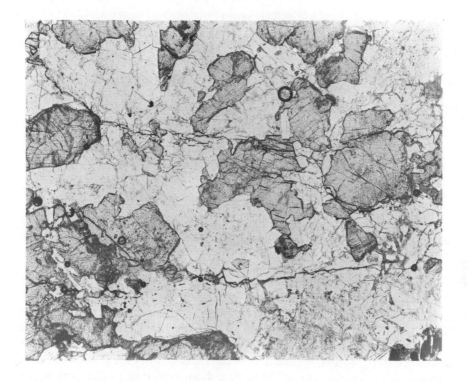

FIGURE 6. Photomicrograph of sample of dolerite from weathered joint blocks shown in Figure 5. Two major cracks are evident. Note how alteration along the cracks is concentrated within the pyroxene crystals (dark areas) and is essentially absent from the feldspar areas (light). Magnification × 48

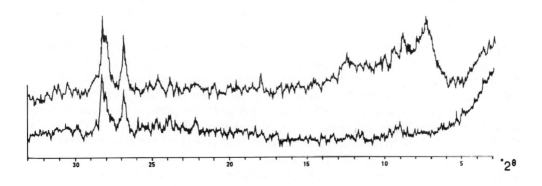

FIGURE 7. X-ray diffraction traces of untreated (upper) and heated (lower) samples of weathered dolerite from rounded joint block of tor. Main peaks are at 28 °2θ(0·318–0·320 nm—plagioclase), 26·8 °2θ (0·34 nm—quartz), 12·4 °2θ (0·72 nm—kaolinite), 8·9 °2θ (1·01 nm—halloysite) and 7·3 °2θ (1·21 nm—probably sepiolite)

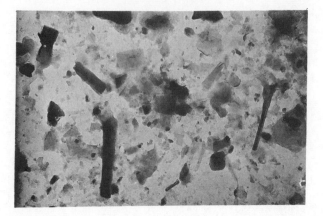

FIGURE 8. Electron photomicrograph of fines from the rounded dolerite joint blocks of the tor. Mixture of quartz and flat and rolled plates of kaolinite (left centre) and with a crystal of halloysite (extreme right). Magnification × 30 000

matical but is closely similar to the *d*-spacing of sepiolite, a hydrated magnesium silicate the conditions of origin of which are not entirely understood (G. Millot, 1970). The disappearance of the peaks on heating the sample to 550°C for two hours, and inspection of the clays under the electron microscope (T. F. Bates, 1955), confirmed the presence of kaolinites. Flat, rolled and partly rolled plates of kaolinite were found to be abundant in some mineral clusters, with minor amounts in the form of halloysite (Fig. 8). These results demonstrate that the rounded joint blocks have been subjected to chemical weathering. This has affected only the ancillary biotite to form kaolinite and the pyroxenes which occasionally show altered margins. The feldspars appear fresh throughout.

The weathering characteristics described above suggest two possible origins for the summit tors. The first possibility is that, while chemical weathering of dolerite may always have occurred in the dry valleys, the rapidity of frost shattering has generally prevented chemically derived residues from being preserved on rock surfaces. Moreover, working along the cracks created by frost action, chemical weathering may have been accelerated in the phases of warmer climate suggested by other evidence, thus tending to produce rounded rock surfaces. In other words, a variable balance between the rate of chemical alteration and the rate of mechanical weathering might produce characteristics such as those displayed by the rounded tors. Given this hypothesis, the summit tors near Sandy Glacier must be regarded as in process of destruction by frost action under the present climate, for all show some frost-shattered faces and some show freshly shattered faces only.

A second possible hypothesis is that chemical weathering has been slow and continuous since at least the end of the Taylor IV glaciation of Denton, Armstrong and Stuiver (1969), i.e. since at least 2·2 million years ago, and that both chemically weathered and unweathered rock outcrops can be expected to occur contiguously as a result of local factors of climate, erosion and mass wasting. According to this hypothesis, chemical weathering is likely to have been most prolonged on ridges having little capacity to accumulate snow sheets and drifts and which, consequently, have been rendered extremely xeric. Certain sites, such as cols or broad passes, are more susceptible than convex spurs and salients to snow accumulation which could be expected to make the microclimate moister and the efficacy of frost shattering greater, thus inhibiting chemical weathering of rock surfaces. It will be noted that the angular summit tors described here are found in a broad col (A in Fig. 2) which, to judge from U.S.A.F. aerial photographs taken in early summer, appears to receive much more snow than the lower part of the arête where the tors become rounded (R in Fig. 2). Thus the distinction between angular and rounded tors appears to be related to the varying susceptibility of different sites to accumulation of drifting snow.

At least on the basis of the limited area examined in the field, therefore, it would appear that the second hypothesis conforms best with the evidence available, for the rounded tors, which occupy dry summit sites, show chemical weathering residues, while elsewhere bedrock surfaces are fresh, frost-shattered and wind-polished. The widespread agreement evident in the literature that chemical weathering on rock surfaces is probably absent in the McMurdo oasis is difficult to reconcile both with the recognition of widespread and currently active chemical weathering in the superficial deposits and with the evidence presented here. It may be that the rock surface sampled by Kelly and Zumberge (1961) is atypical of general conditions in the dry valleys in being very close to the sea and having emerged from it only in relatively recent geological time. It is to be hoped that further work on old weathered outcrops in the Antarctic oases will test the degree to which these apparently conflicting assessments of rock weathering are representative of the area as a whole. If chemical weathering of rock stacks is found to be general in proglacial sites, such as those of the Sandy Glacier arête, then it appears certain that, while changes toward more temperate glacial régimes may have occurred in the past, these have either been very short-lived or have affected the lower parts only of the alpine glaciers. The greater snow accumulation and more widespread snow cover associated with more temperate phases appears likely to have destroyed such chemically weathered rock surfaces.

Evidence such as bouldery ice-cored cirque moraines, and cirque glacier structures strongly suggesting movement by rotational slip at some time in the past (Dort, 1967a), are certainly best explained by a period of more temperate glaciological régime. However, some of the field evidence suggests that these episodes may have been relatively brief and that, overall, aridity has been dominant. Sandy Glacier itself provides some evidence on this point. It has been shown to be composed throughout of alternating fine layers of sand and ice and, as both constituents are therefore likely to have been derived from strong drifting from the west (Dort, E. F. Roots and E. Derbyshire, 1969), the glacier constitutes a record of accumulation under dry, windy conditions throughout its history. While the precise periodicity of sedimentation is unknown, at least some thousands of years of such conditions seem to be indicated. Accordingly, it would appear that the general climate in which the Sandy Glacier tors developed has been one of extreme aridity and that phases of more maritime climate have been relatively slight and short-lived and so have played no significant part in the modelling of the tors. Both the hillslope tors and the angular summit tors, however, reflect local microclimatic conditions different from those prevailing over most of the lower summit area of the arête. The angular summit tors are regularly surrounded by drift snow which accelerates summer frost-shatter. The hillslope tors, while probably receiving little snow cover, face west and north-west and are therefore subjected to relatively more intense heating from the low-angle summer sun. The temperatures of rock surfaces in the Antarctic oases are known to exceed 30°C under the influence of solar radiation.

Finally, it may be that the Sandy Glacier tors have an even wider significance. J.

Palmer and R. A. Neilson (1962) have claimed that tors are basically angular landforms produced by frost acting on bedrock, with edge-rounding resulting from subsequent amelioration of climate, though they allow for some edge-rounding by frost action on lamellar structures (p. 329). D. L. Linton (1955, 1956, and 1964), on the other hand, argued that the tors are 'prepared' by deep weathering principally along bedrock joints to produce sub-surface rounding of corestones ('woolsacks') under a warm temperate to tropical climate prior to their exhumation under periglacial conditions. Neither of these hypotheses appears likely to offer a ready explanation of the features encountered in southern Victoria Land. Moreover, while angular and rounded tors have been found in the same region before, their differences have generally been explained in terms of petrography, structure, topographical location or regional climatic change. Workers in areas as far apart as Europe, Australia and New Zealand have adduced evidence that the tors within a single region may be polygenetic in origin. Thus the summit tors of the Bohemian Highlands (J. Demek, 1964) and north-eastern Tasmania (N. Caine, 1967) have been regarded as two-cycle forms. Adjacent sub-summit tors are said to result from a single, continuing cycle of slope retreat caused by frost action, with mass movement, running water and nivation as ancillary processes.[2] Standing above a surface of cryoplanation, the angular summit tors, together with the flanking hillside tors, are clearly being 'chiselled out of hard rock by frost action' (Linton, 1964, p. 11), and approximate to the descriptions of the 'palaeo-arctic' tors of J. Palmer and J. Radley (1961). The majority of the Sandy Glacier summit tors are, however, rounded by weathering so that the proposal (Palmer and Radley, 1961) that tors of distinctive shape may be identified with periglacial conditions appears untenable. Equally, a preparatory phase of deep weathering seems unlikely in the case of the Sandy Glacier tors which form part of an arête close to 1500 m, the highest level of inundation of the 'First Glaciation' in the Wright valley as estimated by Bull, McKelvey and Webb (1962). It may be that these slopes have suffered no Pleistocene glaciation, only severe and prolonged periglaciation. Sound rock surfaces surround the tors and no deeply weathered deposits were discovered. Moreover, as has been suggested, major expansion even of these cirque glaciers to overrun the arêtes seems unlikely to have occurred during Pleistocene time, if ever.[3] Accordingly, deep weathering cannot be regarded as an essential requirement of tor formation, even in the case of rounded, chemically weathered 'woolsack' tors.

In recognizing geomorphological distinctions in tors within a single massif arising from locally varying formative processes, R. A. Pullan (1959) and Demek (1964), among others, have suggested that a single hypothesis of tor formation cannot be expected to cover all cases within a given area. The evidence presented here fortifies this view and the corollary arising from it (*cf.* Palmer and Neilson, 1962; Caine, 1967) that genetic definitions of tors cannot be universally applied. In particular, the evidence from Sandy Glacier draws attention to the fact that several geomorphological processes may act concomitantly on the same slope to produce a variety of tor morphologies, including not only angular and rounded tors but exfoliating corestones, all under a periglacial climate. This variety may be at least as much a product of local site differences, affecting microclimate, as of regional climatic change.

ACKNOWLEDGEMENTS

Field studies on which this report is based were made possible by the Visiting Foreign Scientists Program, Office of Antarctic Programs, National Science Foundation (U.S.A.), under a research grant (GA-688) awarded to Dr W. Dort,

Department of Geology, University of Kansas. The author records his particular indebtedness to Dr Dort for considerable help and discussion in the field. He also thanks the Departments of Geology and Physics in the University of Keele for help with X-ray diffractometry and electron microscopy.

NOTES

1. This substantiates the opinion of Tedrow and Ugolini (1966) that chemical weathering is widespread in this region.

2. Nivation has played some part in the evolution of the angular hillslope tors at Sandy Glacier, to judge from limited development of protalus ramparts. These are only one boulder thick and lie on a south-west-facing slope (Fig. 3) suggesting an origin by drifting dominantly from between east and north-east.

3. It has been proved that widespread glaciation was established in Antarctica in the late Tertiary (R. H. Rutford, C. Craddock and T. W. Bastien, 1968).

REFERENCES

ANGINO, E. E., M. D. TURNER and E. J. ZELLER (1962) 'Reconnaissance geology of Lower Taylor Valley, Victoria Land, Antarctica', *Bull. geol. Soc. Am.* 73, 1553–62

BATES, T. F. (1955) 'Electron microscopy as a method of identifying clays', *Clays and Clay Technology, Bull. Div. Mines Calif.* 169, 130–50

BELL, R. A. (1966) 'A seismic reconnaissance in the McMurdo Sound region, Antarctica', *J. Glaciol.* 6, 209–21

BULL, C., B. C. MCKELVEY and P. N. WEBB (1962) 'Quaternary glaciations in southern Victoria Land, Antarctica', *J. Glaciol.* 4, 63–78

CAILLEUX, A. (1962) 'Études de géologie au détroit de McMurdo (Antarctique)', *Comité National Français pour les recherches antarctiques*, No. 1

CAINE, N. (1967) 'The tors of Ben Lomond, Tasmania', *Z. Geomorph.* 11, 418–29

CALKIN, P. (1964) 'Geomorphology and glacial geology of the Victoria Valley system, southern Victoria land, Antarctica', *Ohio St. Univ., Inst. Polar Stud. Rep.* No. 10

CAMPBELL, I. B. and G. G. C. CLARIDGE, (1967) 'Site and soil differences in the Brown Hills region of the Darwin Glacier, Antarctica', *N. Z. J. Soil Sci.* 10, 563–77

DEMEK, J. (1964) 'Castle koppies and tors in the Bohemian Highland (Czechoslovakia)', *Biul. peryglac.* 14, 195–216

DENTON, G. H., R. L. ARMSTRONG and M. STUIVER (1969) 'Histoire glaciaire et chronologie de la région du détroit de McMurdo, sud de la Terre Victoria, Antarctide: note préliminaire', *Rev. Géogr. phys. Géol. dyn.* 11, 265–78

DORT, W. (1967) 'Geomorphic studies in southern Victoria Land', *Antarctic J. U. S.* 2, 113

DORT, W. (1967a) 'Internal structure of Sandy Glacier, southern Victoria Land, Antarctica', *J. Glaciol.* 6, 529–40

DORT, W. (1970) 'Climatic causes of alpine glacier fluctuations, southern Victoria Land', *International Symposium on Antarctic glaciological exploration* (ISAGE), *Bull. int. Ass. scient. Hydrol.* 86, 358–62

DORT, W., E. F. ROOTS and E. DERBYSHIRE (1969) 'Firn-ice relationships, Sandy Glacier, Southern Victoria Land, Antarctica', *Geogr. Annlr* 51A, 104–11

EVERETT, K. R. and R. E. BEHLING (1970) 'Chemical and physical characteristics of Meserve Glacier morainal soils, Wright Valley Antarctica: an index of relative age?', *International Symposium on Antarctic glaciological exploration* (ISAGE), *Bull. int. Ass. scient. Hydrol.* 86, 459–60

GUNN, B. M. (1962) 'Differentiation in Ferrar dolerites, Antarctica', *N. Z. J. Geol. Geophys.* 5, 820–63

KELLY, W. C. and J. H. ZUMBERGE (1961) 'Weathering of a quartz diorite at Marble Point, McMurdo Sound, Antarctica', *J. Geol.* 69, 433–46

LINTON, D. L. (1955) 'The problem of tors', *Geogrl J.* 121, 470–87

LINTON, D. L. (1956) 'The significance of tors in glaciated lands', *Proc. 17th int. geogr. Congress* (Washington, 1952), 354–7

LINTON, D. L. (1964) 'The origin of the Pennine tors—an essay in analysis', *Z. Geomorph.* 8 (Sonderheft), 5–24

McGRAW, J. D. (1967) 'Soils of Taylor Dry Valley, Victoria Land, Antarctica, with notes on soils from other localities in Victoria Land', *N. Z. J. Geol. Geophys.* 10, 498–539

McKELVEY, B. C. and P. N. WEBB (1962) 'Geological investigations in southern Victoria Land, Antarctica', *N. Z. J. Geol. Geophys.* 5, 143–62

MILLOT, G. (1970) *The geology of clays*

NICHOLS, R. L. (1963) 'Geologic features demonstrating aridity of McMurdo Sound area, Antarctica', *Am. J. Sci.* 261, 20–31

NICHOLS, R. L. (1965) 'Antarctic interglacial features', *J. Glaciol.* 5, 433–49

NICHOLS, R. L. (1966) 'Geomorphology of Antarctica' in *Antarctic soils and soil forming processes* (ed. J. C. F. TEDROW), 1–46

PALMER, J. and R. A. NEILSON (1962) 'The origin of granite tors on Dartmoor, Devonshire', *Proc. Yorks. geol. Soc.* 33, 315–40

PALMER, J. and J. RADLEY (1961) 'Gritstone tors of the English Pennines', *Z. Geomorph.* 5, 37–52

PÉWÉ, T. L. (1960) 'Multiple glaciation in the McMurdo Sound region, Antarctica—a progress report', *J. Geol.* 68, 498–514

PULLAN, R. A. (1959) 'Tors', *Scott. geogr. Mag.* 75, 51–5

RUTFORD, R. H., C. CRADDOCK and T. W. BASTIEN (1968) 'Late Tertiary glaciation and sea level changes in Antarctica', *Palaeogeogr. Paleoclimatol. Palaeoecol.* 5, 15–39

TEDROW, J. C. F. and F. C. UGOLINI (1966) 'Antarctic soils' in *Antarctic soils and soil forming processes* (ed. J. C. F. TEDROW), 161–77

UGOLINI, F. C. and C. BULL (1965) 'Soil development and glacial events in Antarctica', *Quaternaria* 7, 251–69

United States Navy (1965) *Climatic Atlas of the World, VII—Antarctica*

WILSON, A. T. (1964) 'Evidence from chemical diffusion of a climatic change in the McMurdo dry valleys 1200 years ago', *Nature, Lond.* 201, 176–7

RÉSUMÉ. *Rochers ruiniformes, altération atmosphérique et climat dans le sud de la Terre Victoria (Antarctique)*. Description des traits généraux de la géomorphologie et du climat de « l'oasis » McMurdo, dans le sud de la Terre Victoria (Terre Antarctique). Description morphologique d'un groupe de rochers ruiniformes (« tors ») sur une arête du glacier Sandy (vallée de Wright), et description de leur altération atmosphérique. Ces rochers aux diaclases carrées sont, les uns angulaires, les autres arrondis. La présence de minéraux argileux du groupe de la kaolinite, à la surface et directement sous la surface des blocs arrondis, atteste de l'altération chimique de la diabase. La juxtaposition de deux types de rochers ruiniformes (arrondis et angulaires) au sommet de l'arête s'explique par des variations topographiques et climatiques locales, plus spécialement lorsqu'elles affectent l'épaisseur de la couverture neigeuse. Ces deux types de « tors » résultent des conditions *ultrapériglaciaires* prédominantes. On peut conclure que ces conditions ont probablement dominé pendant tout le Quaternaire, avec seulement de brèves périodes de climat légèrement plus maritime, et que les variations dans la morphologie des « tors » résultent au moins autant des différences d'emplacement qui influent sur le micro-climat, que des changements climatiques sur l'ensemble de la région.

FIG. 1. Carte de localisation montrant les vallées sèches du sud de la Terre Victoria. Le glacier Sandy se trouve à l'ouest-nord-ouest de la base de McMurdo, Île de Ross

FIG. 2. Mosaïque de photographies aériennes montrant le glacier Sandy et, à l'est, une partie du glacier sans nom. La localisation des rochers ruiniformes de sommet est indiquée (A : angulaires ; R : arrondis). On voit nettement la localisation des corniches de versant formées de grès Beacon (B), de même que les moraines de cirque à noyau de glace (MM). L'orientation est approximativement nord-sud (Photographies de l'armée de l'air des États-Unis)

FIG. 3. Corniches de versant formées de grès Beacon sur les premières pentes de l'arête du glacier Sandy (B dans la figure 2). L'alignement en croissant de rochers de diabase (premier plan) est une moraine de névé provenant de corniches angulaires de diabase (à l'extérieur de la photographie, côté gauche). L'échelle est indiquée par la silhouette humaine au deuxième plan, à gauche du centre. Vue vers le sud-est

FIG. 4. Pinacle angulaire de diabase Ferrar modelé par la gélivation. À l'arrière plan, des polygones macmurdiens, une moraine à noyau de glace et le cirque du glacier sans nom. Vue vers le nord

FIG. 5. Pinacle arrondi de diabase (du type « woolsack »). À gauche écailles d'exfoliation, à droite, une face gélivée. On a prélevé des échantillons de cette corniche pour en étudier la minéralogie et l'altération chimique

FIG. 6. Microphotographie d'un échantillon de diabase provenant des rochers aux diaclases carrées (voir figure 7). Deux fissures importantes sont nettement visibles. Les produits de l'altération chimique le long des fissures sont concentrés dans les cristaux de pyroxène (aires obscures) ; les feldspaths, au contraire, sont en majeure partie, intacts (aires pâles). Grossissement × 48

FIG. 7. Traces diffractométriques des rayons X de quelques échantillons de diabase altéré provenant d'un bloc arrondi de la corniche de la figure 5 : sans traitement (en haut), et après chauffage (en bas). Les principaux points culminants atteignent à 28 $^\circ\theta$ (0,378–0,320 nm—plagioclase), 26.8 $^\circ 2\theta$ (0,34 nm—quartz), 12.4 $^\circ 2\theta$ (0,72 nm—kaolinite), 8.9 $^\circ 2\theta$ (1,01 nm-halloysite), 7.3 $^\circ 2\theta$ (1,21 nm-probablement sépiolite)

FIG. 8. Microphotographie à électrons montrant les argiles et des limons provenant d'un bloc arrondi de la corniche de la figure 5. Mélange de quartz et de kaolinite en plaques plates et bouclées (à gauche du centre), et avec un cristal d'halloysite (à l'extrême droite). Grossissement × 30 000

ZUSAMMENFASSUNG. *Felsburgen, Verwitterung und Klima auf Sudviktorialand, Antarktis*. Es werden die allgemeinen geomorphologischen und klimatischen Merkmale der Mc.Murdo, ,Oase', Sudviktorialand, bescrieben. Es werden morphologische Wesenszüge und Verwitterungsmerkmale einer Gruppe von Felsburgen, (,tors') am Sandy Gletscher (Wright-Tal) behandelt. Es kommen sowohe eckige als auch gerundete geklüftete Blöcke vor. Das Vorhandenseim von tonhältigen Gesteinen (Kaolinit-Gruppe) auf und innerhalb der Oberfläche der gerundeten Kernsteine (unter der Oberfläche verwittert) weist auf chemische Verwitterung des Dolerit hin. Das Nebeneinander von gerundeten und eckigen Felsburgen auf dem Gratgipfel lässt such durch die ötlichen Unterschiede der topographischen und klimatischen Bedingungen erklären, vor allem deswegen, weil sie die schneedecke beeinflussen. Beide Felsburgen typen sind das Produkt vorherrschend polarischer Wüstenbedingungen. Es wird der Šchluss gezogen, dass solche Bedingungen wahrscheinlich im ganzen Pleistozän vorgeherrscht habenmit nur kurzen Abschnitten eines mehr ozeanischen

Klimas und dass die Vielfalt in der Felsburgen-Morphologie in zumindest gleichem Masse das Ergebnis von örtlichen Lageunterschieden, die das Kleinklima beeinflussen, wie von regionalem Klimawechsel ist.

ABB. 1. Orientierungskarte der Trockentäler, Südviktorialand. Der Sandy Gletscher liegt 100 km nord-nordwestlich der McMurdo Station, Insel Ross

ABB. 2. Unkontrolliertes Luftbildmosaik, das den Sandy Gletscher und einen Teil des namenlosen Gletschers gegen Osten zeigt. Es werden die Standorte der eckigen (A) und gerundeten (R) Gipfelfelsburgen und Berhangfelsburgen in der Beacon Sandsteinreihe (B) gezeigt, sowohl als auch die Kargletschermoränen mit Eiskern (MM). Orientierung ungefähr Nord zu Süd. Bilder der U.S.A. Luftwaffe

ABB. 3. Berghangfelsburgen in der Beacon Sandsteinreihe auf den unteren Hängen des Sandy Gletschergrates (B in Abb. 2). Die halbmondförmige Anordnung der Doleritblöcke (Vordergrund) ist eine Firnmoräne, die auf eckige Doleritfelsburgen zurückgeht (links vom Bild). Die Zahl im Zentrum links mittlere Entfernung gibt den Massstab an Blick gegen Südosten

ABB. 4. Eckige, frostzerstörte Felsburg aus Ferrar Dolerit mit Spaltenfrostpolygonen, Moräne mit Eiskern und namenloser Kargletscher im Hintergrund. Blick gegen Norden

ABB. 5. Gerundete (,Woolsack') Felsburg aus Dolerit, die Abschuppungs (Dequamations) schalen (links) und frostzerstörte wand (rechts) zeigt. Proben für Mineralogie- und Verwitterungsforschung wurden von dieser Felsburg genommen

ABB. 6. Photomikrograph einer Doleritprobe von Verwitterten, durch Klüfte verbundenen Blöcken, gezeigt in Abb. 7. Man kann zwei Hauptsprünge sehen. Man beachte, wie sich die Veränderung entlang der Sprünge auf die pyroxenen Kristalle (dunkel Stellen) konzentriert und im wesentlichen an den Feldspatstellen (hell) nicht vorkommt. Vergrösserung $\times 48$

ABB. 7. Kurvendurstellung der Beugung von Röntgenstrahlen von nicht behandelten (oberen) und erhitzten (unteren) Probem eines verwitterten Dolerits, der von einem, gerundeten geklüfteten Felsburgblock stammt. Höchstpunkte sind bei 28 °2θ (0,318–0,320 nm—Plagioklas), 26.8 °2θ (0,34 nm—quartz), 12.4 °2θ (0,72 nm—kaolinit), 8.9 °2θ (1,01 nm—Halloysit) und 7.3 °2θ (1,21 nm—coahr-scheinlich Sepiolit)

ABB. 8. Elektronenphotomikrograph von Ton und Schlick eines gerundeten, geklüfteten Felsburgblockes. Mischung aus quartz und flachen und gerollten Kaolinitplatten (Mitte links) und mit einem Halloysit Kristall (ganz rechts). Vergrösserung $\times 30\ 000$

Valley asymmetry and slope forms of a permafrost area in the Northwest Territories, Canada[1]

BARBARA A.KENNEDY

(*Demonstrator in Geography, Cambridge University*)

AND MARK A.MELTON

(*Visiting Associate Professor of Geography,
Simon Fraser University*)

Revised MS received 8 October 1971

ABSTRACT. Valley asymmetry and slope forms in a small area of sedimentary rocks adjacent to the Mackenzie River delta are examined in relation to the major variations in climate, available relief and current processes. The study area is very varied geomorphologically and it does not fit the classic 'periglacial' pattern. Asymmetry in maximum slope angles is analysed statistically and found to reverse between (a) areas of more severe climate and low available relief—where north-facing slopes are significantly steeper than those facing south—and (b) the zone of milder climate and deeper valleys, where south-facing slopes are the steeper. Slopes that are directly under the control of fluvial erosion show less response to differences in the degree of basal corrasion than to variations in aspect, a finding which differs from the results of studies in non-permafrost areas.

DESPITE a recent surge of geomorphological interest in high-latitude areas, there is, at the present time, a notable scarcity of accurate quantitative information concerning the basic features of stream valleys in such regions. This lack of data hinders understanding of valley forms in zones of Pleistocene permafrost, currently experiencing more temperate climates. The work presented here consists of a quantitative study of slope forms in a small area of contemporary permafrost, underlain by weakly consolidated sedimentary rocks. An attempt has been made to relate the most notable variations in form—and particularly the asymmetry in slope angle between north- and south-facing profiles—to what appear to be the major differences in the contemporary erosional environment.

LOCATION AND NATURE OF THE STUDY AREA

Field investigations were carried out in an area of 9 km² in the Caribou Hills, Northwest Territories, Canada, centred on the hamlet of Reindeer Station (134° 71'W; 68°42'N; Figs. 1 and 2). The hills in this area rise directly from the East Channel of the Mackenzie River to a maximum elevation of just over 180 m and are underlain by alternating beds of fine greyish-white sands and white claystones. The two lithologies form distinct relief units on south-facing slopes in the lower valleys (shown in Fig. 4), where short cliff sections developed on the sands ('risers') are separated by lower-angle claystone 'treads'. The apparent dip of the beds is not more than 1° to the north-west and there is no evidence of folding or faulting.

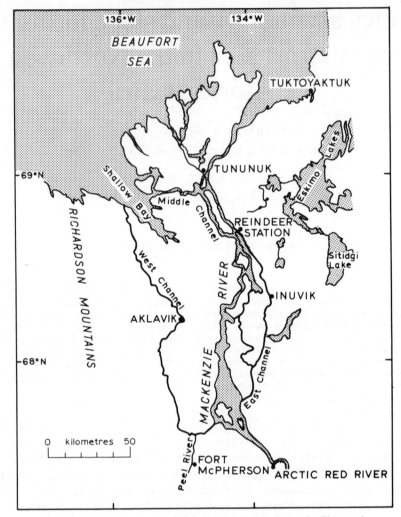

FIGURE 1. The Mackenzie delta region (after J. R. Mackay, 1963, Figure 71)

Ice retreat in the area is thought to have commenced in about 12 000 BP, but major glacial and fluvio-glacial forms are absent (J. R. Mackay, 1963, p. 21 and 27) and it seems possible that the present west-draining valleys largely post-date this retreat, as the major divide is still very close to East Channel and the valley forms are youthful (Figs. 2 and 3).

The history of base-level changes since 12 000 BP is rather complex (Mackay, pp. 38–43), but the most recent major event is a relative rise of the order of 3–6 m that has resulted in infilling of the lower reaches of the west-trending valleys by rather high-angle (6–13°) alluvial fans, on which stream courses are braided. A slight negative movement may have occurred in the last few years, as some trenching of the fan surfaces has taken place and many channels in the lower valleys are consequently rather fixed in position. Evidence of the aggradational episode does not extend to the upper valleys of the scarp front, nor to the plateau top, where low terraces occur at 1–3 m above present water levels.

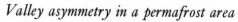

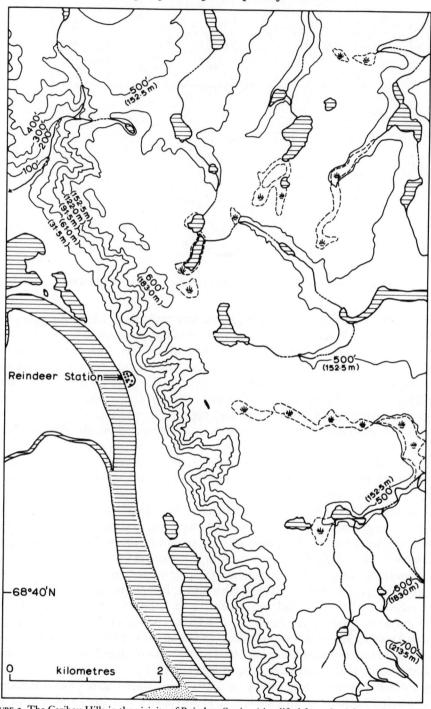

FIGURE 2. The Caribou Hills in the vicinity of Reindeer Station (simplified from the Advance Print of sheet 107B/11E, 1:50 000 series, Surveys and Mapping Branch, Department of Mines and Technical Surveys, Ottawa)

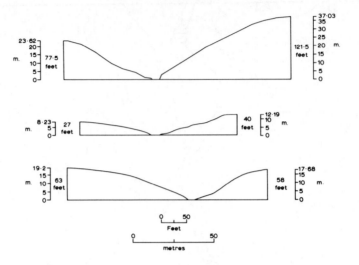

FIGURE 3. Some examples of cross-sections in the upper valleys, Caribou Hills

The climate of the field area is defined generally by Mackay (ibid., p. 153) as transitional between the arctic and sub-arctic and there is in fact progression from arctic to sub-arctic as one moves from the plateau top to the base of the hills. Apart from the generally severe temperature régime (annual extremes about $+27°$ to $-45°C$), the most important climatic feature is the very low total annual precipitation (*c.* 250 mm). Although snowfall is light, drifting occurs and some snow patches on north-east-facing slopes were observed to be almost 1 m deep as late as the end of June, 1967.

Low-order streamflow in the area appears more dependent on the thermal than the precipitation régime, as most moisture is derived from snowmelt and the thawing of the active layer of the permafrost. There is, however, a summer maximum of precipitation and this appears to be important in maintaining flow in higher-order channels. All stream flow is best classed as intermittent.

Those westward-flowing streams investigated were found to flow on steep slopes (4–22°) and to be generally incised, with beds of clayey silt or organic muck, reflecting the generally fine-grained nature of the bedrock. It is assumed that most material reaching the channels must be carried largely by suspension or solution and the lack of turbidity noted in summer appears to suggest that maximum transport of sediment occurs in spring (*cf.* Mackay, ibid., p. 43; L. Arnborg, H. J. Walker and J. Peippo, 1967).

The area is covered with soil and vegetation and the vegetation reflects the major variations in climate. Spruce (*Picea glauca, P. mariana*) is restricted to south-facing slopes in the deep lower valleys, while large willows (*Salix* sp.) occur on the fan surfaces at the valley mouths and are replaced, upstream, by thick growths of alder (*Alnus crispa*). In other sites, the vegetation is dominated by ground birch (*Betula glandulosa*) and dwarf willows (*Salix* sp.) which grow most successfully on south-facing slopes. Bare ground is restricted to a few free-face sections on the sands and to scar zones associated with recent bank-caving or mud-flows.

Although the whole area is underlain by continuous permafrost, with an active layer only 0·1–0·6 m in depth, the completeness of the cover of vegetation tends to suppress signs of 'normal' periglacial processes, particularly solifluction (*cf.* J. B. Bird, 1953; Mackay, 1958; R. Thorsteinsson and E. T. Tozer, 1962). It must, however, be assumed that some soil movement as a result of ice segregation will occur throughout these valleys. Signs of other processes that were noted in the area are listed in Table I.

TABLE I

Nature and location of signs of geomorphological processes observed in the Caribou Hills, 1967

A	Type	Site
'Periglacial'	1. Tundra polygons: high centre	Old lake beds, plateau top
processes	2. Turf-hummocks, ≤ 300 mm high	Particularly north-facing slopes, upper valleys
	3. Vegetation stripes, 1·5–2·0 m wide	Particularly low-angle south- and west-facing profiles
	4. Snow-patch hollows, vegetated	North- and east-facing slopes
B		
'Normal'	1. 'Wash-and-creep' (*cf.* S. A. Schumm, 1964)	Upper 50 mm of bare claystones on interfluves
processes	2. Deflation	Sands, on interfluves
	3. Mudflows	Main valleys and scarp front: no preferred orientation
	4. Slumps and landslips	North-facing, basally corraded slopes
	5. Minor sliding	North-facing slopes, downhill from turf hummocks
	6. Terracettes	Poorly vegetated south-facing slopes, lower valleys
	7. Gullying	South-facing slopes, lower valleys

Little need be said of any of these features, with the exception of snow-patch hollows, slumps and landslips, and gullying.

(1) The location and extent of the vegetated snow-patch hollows common on north-facing slopes is of interest. On the plateau top, these features occur at the base of profiles and may occupy as much as 50 per cent of their length; in the upper valleys, the site is similar, but the proportion of the profile involved falls to 25 per cent or less; in the lower valleys, the hollows are found near the heads of slopes and rarely account for as much as 10 per cent of the total length of slope (Fig. 4).

(2) The most common type of slump and landslip is massive bank-caving, restricted to north-facing profiles: that is, to those slopes with the permafrost table within perhaps 0·1 m of the surface, even in summer. The location of these features conflicts with the assumption frequently made (e.g., H. T. U. Smith, 1949) that deeply frozen slopes are immune to basal attack by streams. Clearly, there will be substantial differences between slopes with shallow and with deep active layers: the most important difference in this context may be the degree of consolidation of the surface material. One can perhaps envisage rather pronounced lateral erosion and undercutting of profiles in the former class before failure along the permafrost plane occurs; whereas in the latter case there will be less resistance and, therefore, earlier and less wholesale slumping and plastic flow of the bank materials into the channel.

(3) The gullying of south-facing walls in the lower sections of valleys is extremely pronounced. It is assumed that the location and extent of these features is closely associated with the relatively great depth of the active layer on these sites, in that this will both provide a large volume of surface run-off in spring and early summer and, in high summer, lead to the creation of a semi-arid microclimate that inhibits the growth of vegetation.

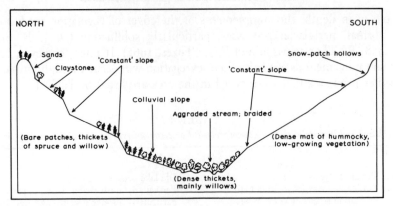

FIGURE 4. Diagrammatic cross-profile of a lower valley section, Caribou Hills

In summary, this small area is one that exhibits an unusual complexity of geomorphological environments yet, although underlain by continuous permafrost, it is not immediately recognizable as a 'periglacial' landscape.

EXPERIMENTAL DESIGN AND FIELD TECHNIQUES

There are three major controls of slope form within the area:

(1) **Topographical position** The plateau top, upper and lower valley sections form distinct units on the basis both of immediate relief (less than 10 m; about 20 m; over 90 m) and of climate (arctic; transitional; sub-arctic).

(2) **Aspect** Widespread variation in angle of slope between north- and south-facing profiles has been noted for many areas of past and present permafrost (e.g., D. R. Currey, 1964; J. Grimberieux, 1955; D. M. Hopkins and B. Taber, 1962; S. Judson and G. W. Andrews, 1955; M. P. Kerney, E. H. Brown and T. J. Chandler, 1964; J. N. Malaurie, 1952).

(3) **Local erosional environment** The presence or absence of basal undercutting by a stream channel is likely to produce marked variation in the steepness of slip-off and undercut profiles (A. N. Strahler, 1950; M. A. Melton, 1960).

It was decided to investigate primarily the influence of factors (1) and (2) and to carry out a more complete analysis of factors (2) and (3) in the upper valley sections, where streams are both present and 'meandering'. The general absence of east-facing slopes in this zone (Fig. 2) prevented observations being made on slopes of all four cardinal aspects, but samples of west-facing profiles were obtained in addition to those of north- and south-facing slopes. The experimental design is shown in Figure 5 and was set up to allow the use of factorial analysis of variance (E. L. Leclerg, W. H. Leonard and A. G. Clark, 1966). Three levels of analysis have been employed:

(1) The effects of topographical position and aspect upon three features of slope form (maximum angle; length of steepest section; degree of vegetation cover on the steepest section) are estimated by a two-way factorial experiment, employing three levels of topographical position (plateau top; upper valley; lower valley) and two of aspect (north-facing; south-facing), giving six 'treatments' in all.

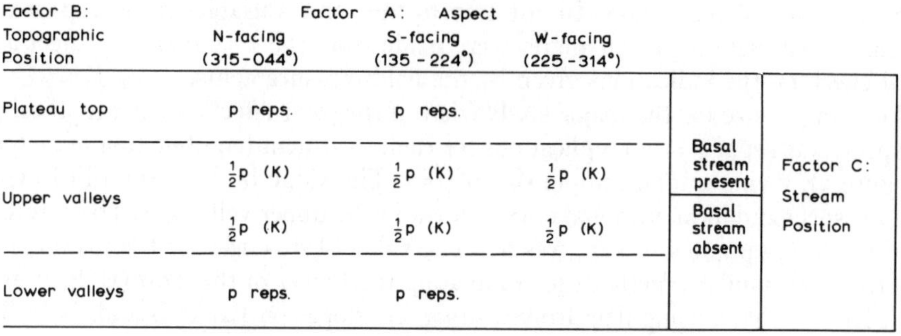

Factor B: Topographic Position	Factor A: Aspect				
	N-facing (315–044°)	S-facing (135–224°)	W-facing (225–314°)		
Plateau top	p reps.	p reps.			
Upper valleys	½p (K)	½p (K)	½p (K)	Basal stream present	Factor C: Stream Position
	½p (K)	½p (K)	½p (K)	Basal stream absent	
Lower valleys	p reps.	p reps.			

FIGURE 5. Complete experimental design, Caribou Hills

(2) The effects of aspect alone on other variables in the plateau-top and lower-valley sites have been assessed by one-way analysis of variance.

(3) The effects of aspect and local erosional environment (Melton, 1960) in the upper valleys are estimated by a two-way factorial design employing three levels of aspect (north-; south-; and west-facing) and two of local erosional environment (presence and absence of a stream undercutting the slope).

There are substantial differences in the geomorphological environments between the three classes of topographical site and this necessitated different sampling procedures in each one. First, on the plateau top, well-defined slope forms are confined to the margins of decaying lakes, old lake-beds and former stream channels. These were sampled by initiating a series of profiles at that point where the aspect of the slope first fell within the azimuthal range 315–044° or 135–224°. Subsequent profiles were spaced along the slope at distances equal to one-half the length of the preceding profile: it is hoped that this imparts the necessary random element to the sampling. The form of each slope was measured by Abney level and tape (accurate to ½° and 0·3 m respectively) and the degree of vegetation cover estimated by eye. The following variables were assessed at these sites: maximum angle of slope; mean angle of slope (height/ground length ratio); profile height; profile length; length of steepest section; vegetation cover of steepest section; weighted average vegetation cover of the whole profile.

Secondly, in the upper valleys, where well-defined valleys exist, sampling was initiated at the upstream end of each, at the first marked bend, and subsequent samples were located at each well-developed sinuosity until the stream debouched into the main valley. Profiles were measured in pairs, and the aspect and local erosional environment of each were noted. The variables measured were similar to those for the plateau-top sites, with the distinction that the maximum angle here had to be defined as the steepest non-cliffed section of the slope.

Thirdly, the lower valleys presented difficulties of access, particularly on their northern sides, and these profiles are also highly dissected. It proved impossible to survey true cross-profiles and the solution adopted was to measure angle, length and vegetation cover on 'constant slope' segments of both classes of valley side. In the case of the south-facing slopes, these segments were taken as the lowest accessible risers on spur-ends. It is possible that this sampling procedure may have introduced considerable statistical bias into the measurements, but none of the distributions produced was found to depart from the expected normal distribution and it is suggested that they may be taken, with reservations, as representative of these sites. In addition to the three variables measured in the field, independent estimates of mean angles were made from the 1:50 000 map sheet as a very general check on the indications given by the field measurements.

The sample size for the major study (that of the joint effect of topographical position and aspect) was set at $q = 30$ replications for each treatment (combination of factor levels: see Figure 5), giving a total sample size of 180. The value of k for the third experiment (effect of aspect and local erosional environment in the upper valleys) was therefore fixed at 15 and the total sample size here was 90. The value of $n = 180$ for the main experiment was derived by standard methods for evaluating the Power of the Test (E. S. Pearson and H. O. Hartley, 1951), using data from comparable slopes on Banks Island, N.W.T.[2]

VARIATIONS IN SLOPE FORMS

With topographical position and aspect

All three variables—the degree, length and vegetation cover of steepest sections—show a highly significant interaction ($p < 0.001$) between topographical position and aspect (Fig. 6, A, B & C). That is to say, in no case is it possible to predict the average response in any one type of site from the simple addition of the mean effects of each factor. Such interactions are undoubtedly common in geomorphological situations, but have not received much attention to date. This is unfortunate, as they provide clear indications of 'anomalous' responses produced by certain combinations of controlling factors.

The variation in each of the three variables will be considered in turn:

Maximum angle Duncan's Test (D. B. Duncan, 1955) was used to estimate the significance of differences between all possible pairs of the six means. The results indicate that, at each topographical site, there is a significant difference ($p < 0.05$) between the mean maximum angles of north- and south-facing profiles. In addition, the values for the plateau-top locations are significantly lower than those from the scarp-front valleys. The source of the significant interaction lies in the 'crossing-over' of the steeper aspect levels between the upper and lower valleys (Fig. 6A): the values for north-facing slopes at the former group of sites and south-facing at the latter do not differ significantly; nor do those of south-facing (upper valley) and north-facing (lower valley) locations.

There is thus clear indication of well-developed asymmetry throughout the area, but also a sharp reversal in the aspect of steeper slopes between the headward and lower sections of the scarp-front valleys. In the zones of the more extreme climate and lesser available relief, north-facing slopes are the steeper, by between 5 and 6°. In the area of sub-arctic climate, and much deeper valleys, south-facing slopes are the steeper by, on average, 5°. This reversal in asymmetry between the upper and lower valley sections appears to be associated with: (1) a significant decline (from 26° to 23°) in the average maximum angle of north-facing profiles; (2) a significant increase (from 22° to 28°) in that of south-facing slopes. Clearly, in an area of such geomorphological complexity, it would be inadvisable to ascribe the pattern of asymmetry to the differential operation of one control. It is suggested that the following is a possible explanation:

(1) A major control of the relative steepness of north-facing profiles is the degree of development of snow-patch hollows. Where these occur near the base of the slope and occupy a high proportion of the total length (as on the plateau top and in the upper valleys),

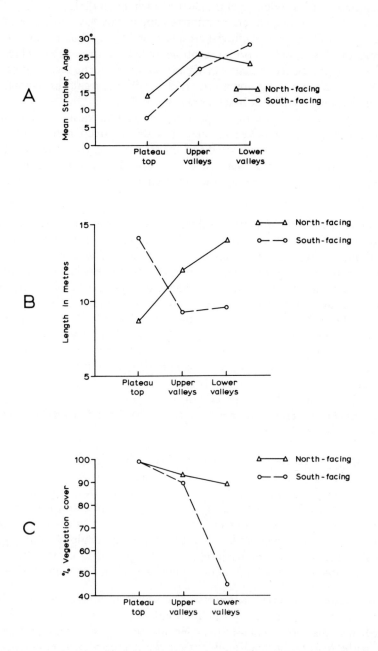

FIGURE 6. Interaction of the effects of aspect and topographical position: A, average Strahler angles; B, average length of Strahler-angle sections; C, average percent vegetation cover of Strahler-angle sections

they are associated with steepening of the slope as a whole. In the lower valleys these features occur near the heads of much larger profiles and apparently have little effect (*cf.* Fig. 4).

(2) In the upper valleys, modern fluvial erosion is active, basal undercutting frequent and north-facing profiles appear more susceptible to attack than those which face south, judging from the prevalence of bank-caving and slumping at the former sites. In the lower valleys, streams are central, aggrading and braided, with the result that neither valley side is currently undergoing significant basal corrasion.

(3) The processes on south-facing slopes are, in general, weakly active and associated primarily with the formation of vegetation stripes. On the steep, high, well-lit slopes of the lower valley sections, however, rill wash and apparently recently-initiated gullying seem to be prominent and such processes are generally associated with the maintenance of steep angles (Schumm, 1956). It is clear, however, that prolonged gullying should result in a decline of the slope angle.

It is suggested that asymmetry develops as follows: snow-patches are initiated at the base of low north-facing slopes; nivation hollows develop and the profiles become steeper relative to those facing south; streams become incised and slopes increase in height and angle, but preferential sapping at the base of north-facing profiles causes a further relative increase in their angle; streams continue to downcut and slopes become steeper, though to a lesser extent on north-facing profiles as length of slope increases and the importance of snow-patch hollows declines; base level rises and valley floors aggrade; long north-facing slopes decline through creep; south-facing profiles remain relatively steep as a result of rill wash and shallow gully development. In other words, it is only possible to explain the observed asymmetry by considering the simultaneous variations through time of both microclimate and relief as these apparently affect the balance of geomorphological processes. It is suggested, in the light of this evidence, that many of the theories of 'periglacial' asymmetry that have been propounded rely far too heavily upon variations in the strength of *one* process or control alone.

Length of steepest section The results of this analysis (Fig. 6B) show the very different sequence of 'constant' slope development on north- and south-facing valley sides as available relief increases. The length of this section of north-facing slopes rises steadily, down-scarp; whereas that of south-facing slopes shows a tendency to decline. In itself, this picture is relevant to the behaviour of the maximum angles as there is a general inverse correlation (on any one lithology) between the length and angle of slope segments (Kennedy, 1969). Long, steep slopes are inherently less stable than short sections of the same angle and are liable to deep-seated failure, giving longer, lower slope sections.

Vegetation cover of steepest sections The source of interaction here (Fig. 6C) is the marked reduction of vegetation cover on the south-facing lower valley slopes and it should be recalled that this was essentially included in the original sampling design. More interesting, perhaps, is the general tendency for the degree of vegetation cover to decline, down-scarp, as relative relief and slope gradients increase together with valley width and the amount of direct insolation received.

In all, then, this stage of the investigation produces a clear picture of complexity in geomorphological environments in this small permafrost area and indicates strongly the joint and not altogether predictable effect of variations in both relative relief and microclimate upon several key features of slope form and, consequently, upon valley asymmetry.

Variations with aspect, by topographical position

Only the plateau top and lower valley sites will be considered here, as the effect of aspect was treated simultaneously with that of the local erosional environment in the upper valleys. The three variables used in the main experiment have not been analysed again here.

Plateau tops   Table II lists the response of the four remaining variables to variation in aspect as shown by the one-way analysis of variance.

TABLE II

Effect of aspect on features of slope form

| | Plateau-top sites | | |
Variable	Average value, north-facing slopes	Average value, south-facing slopes	F level and significance
Mean angle	7·1°	5·1°	12·24 $p < 0.001$
Profile height	6·4 m	7·0 m	0·35 $p > 0.05$
Profile length	51·8 m	79·3 m	11·53 $p < 0.01$
Average per cent vegetation cover	99%	98%	4·49 $p < 0.05$

Clearly there is a pronounced difference between north- and south-facing slopes in this area except, critically, with respect to their height. This finding strengthens the suggestion that the asymmetry between the two classes of profiles arises from the differential operation of a modern process—in this case, nivation—and is not merely the reflection of inherited differences in geometry.

Lower valleys   The only remaining variable analysed for these sites was mean angle of slope, obtained from the topographical map. Two samples—of fifteen measurements from south-facing slopes and twenty-three from north-facing ones—were tested by one-way analysis of variance and showed a highly significant difference ($p > 0.001$) between an average value of 18° in the former case and 17° in the latter. This result indicates that the asymmetry in maximum angles, although obtained by a non-random sampling procedure, is really representative of the situation in these valley sections.

Variation with aspect and local erosional environment

Table III illustrates the response of average maximum angles in the upper valleys to the joint operation of variations in aspect and local erosional environment. Only the variation with aspect is found to be significant by the two-way analysis of variance ($p = 0.01 > 0.001$) and Duncan's test indicates that north-facing profiles are significantly steeper than those facing south and west, both of which categories are statistically similar. Several features of this analysis are interesting.

(a) The low effect of variations in local erosional environment is quite contrary to Melton's (1960) findings for a marginally sub-humid non-permafrost area (the Laramie Mountains,

Wyoming) and perhaps illustrates the proposition that stream activity—especially lateral erosion—in permafrost areas is of relatively small importance.

(b) The similarity in form between south- and west-facing profiles underlines the general assumption that slopes of these two aspects have rather comparable micro-climatic régimes, particularly with respect to receipts of direct insolation (*cf.* W. D. Sellers, 1965; Figure 13).

TABLE III

Maximum angles, upper valleys: average values for each combination of aspect and local erosional environment, together with average effects

Aspect	Local erosional environment		Mean	Effect
	Basal stream present	Basal stream absent		
North-facing	27·3°	24·1°	25·7°	+3·1°
South-facing	21·6°	21·9°	21·7°	−0·9°
West-facing	22·4°	18·3°	20·4°	−2·2°
Mean	23·8°	21·4°		$\bar{X} = 22 \cdot 6°$
Effect	+1·2°	−1·2°		$n = 90$

(c) Table III indicates a tendency for north-facing slopes with a stream close to their base to be rather steep: although this is not sufficiently strong to produce a significant interaction term in the analysis, it probably bears out the assumption that such slopes are, to some degree, unusually responsive to basal sapping.

Table IV illustrates the results of the analyses of the remaining six variables, which exhibit a rather low level of response to both controls, but in this case the direct importance of aspect appears moderately subordinate to that of local erosional environment. This is particularly true of the variation in mean angles and the difference between this pattern and that shown by the maximum angles leads one to conclude that the original assumption that fluvial erosion and processes governed by microclimate are jointly important in these valleys was probably correct.

This assumption is strengthened by the interaction of the two controls in the cases of

TABLE IV

Other features of slope form, upper valleys: average values for each aspect and for each category of local erosional environment

Factor levels	Mean angle	Profile height	Profile length	Length of steepest slope	Per cent vegetation cover on steepest slope	Average per cent vegetation
North-facing	16·9°	17·5 m	54·3 m	12·2 m	93	93
South-facing	15·7°	12·6 m	47·9 m	9·1 m	90	90
West-facing	13·9°	10·7 m	46·6 m	10·1 m	94	93
Stream toward	16·8°*	15·0 m	50·6 m	11·0 m	93	93
Stream away	14·9°*	12·2 m	48·5 m	9·8 m	92	91
Overall mean	15·5°	13·5 m†	49·7 m†	10·4 m	92·5	92

* Difference significant, $p = 0 \cdot 05 > 0 \cdot 001$
† Interaction significant, $p = 0 \cdot 01 > 0 \cdot 001$

profile height and length, arising from the unusually large dimensions of undercut, north-facing slopes. There is here an obvious connection with the high value of average maximum angle in this type of location (Table III), but it seems debatable whether one is justified in concluding (*cf.* J. Alexandre, 1958) that the steep angles are merely reflections of the slope lengths. It may well be that angle and length are both the result of some unusual combination of controls in north-facing undercut sites, but it is impossible, on the present evidence, to surmise what this may be.

CONCLUSION

This study is one of a small area within the zone of present-day permafrost and, clearly, it cannot be held to represent conditions throughout this climatic region as a whole. It is hoped, however, that it illustrates three points:

(1) That, far from introducing a simplifying element into landscape development, the presence of permafrost in an area of marked relief, considerable vegetation cover and marginal arctic climate merely adds to the complexity and variety of present-day processes. As many areas which experienced permafrost conditions in the Pleistocene must similarly have retained accented relief and a considerable cover of soil and vegetation (*cf.* R. F. Flint, 1957), together with a measure of fluvial activity, explanations of the 'relict' landforms in such regions must look beyond the operation of solifluction on bare slopes. There is clearly no one suite of permafrost or periglacial landforms.

(2) Similarly, that there is no one form of asymmetry characteristic of permafrost areas. This conclusion has been hinted at by the variety of opinions in the literature already quoted, but it is thought that this is the first study to demonstrate the co-existence of classes of valleys exhibiting steeper north- and south-facing profiles within one small area, where it seems reasonable to suppose that both types of asymmetry are in tune with modern conditions.

(3) That in an area of complex permafrost geomorphology such as the Caribou Hills, the use of investigations based upon analysis of variance may be most helpful in isolating 'regions' of similar and dissimilar morphological characteristics. Clearly an adequate explanation of the observed pattern of variation requires more detailed studies in similar environments. We hope that the present work may indicate areas in which such research could profitably be concentrated.

ACKNOWLEDGEMENTS

This project was financed by the U.S. Army Research Office, Durham, N.C., under grant DA-ARO-D-31-124-G939, project 7360-EN. The grant was obtained through the head office of the Arctic Research Institute, Montreal. We thank Miss J. E. Towler, who assisted in the field, Messrs D. and S. Gill, C. P. Lewis and R. Best for aid with transport, Mr S. Johannssen for arranging accommodation and the Staff of the Arctic Research Laboratory, Inuvik, for their assistance in every way.

REFERENCES

ALEXANDRE, J. (1958) 'Le modelé quaternaire de l'Ardenne Centrale', *Annls Soc. géol. Belg.* 81, 213–332

ARNBORG, L., H. J. WALKER and J. PEIPPO (1967) 'Suspended load in the Colville River, Alaska, 1962', *Geogr. Annlr* 49, 131–44

BIRD, J. B. (1953) 'Southampton Island', *Mem. geogr. Brch Can.* 1

BIRD, J. B. (1967) *The physiography of Arctic Canada*

CURREY, D. R. (1964) 'A preliminary study of valley asymmetry in the Ogotoruk Creek area, northwestern Alaska', *Arctic* 17, 84–98

DUNCAN, D. B. (1955) 'Multiple range and multiple F tests', *Biometrics* 11, 1–42

FLINT, R. F. (1957) *Glacial and Pleistocene geology*

GRIMBERIEUX, J. (1955) 'Origine et asymétrie des vallées sèches de la Hesbaye', *Annls Soc. géol. Belg.* 78, 267–86

HOPKINS, D. M. and B. TABER (1962) 'Asymmetrical valleys in central Alaska', *Spec. Pap. geol. Soc. Am.* 68

JUDSON, S. and G. W. ANDREWS (1955) 'Pattern and form of some valleys in the Driftless Area, Wisconsin', *J. Geol.* 63, 328–36

KENNEDY, B. A. (1969) 'Studies of erosional valley-side asymmetry', Unpubl. Ph.D. thesis, Univ. of Cambridge

KERNEY, M. P., E. H. BROWN and T. J. CHANDLER (1964) 'The Late-glacial and Post-glacial history of the Chalk escarpment near Brook, Kent', *Phil. Trans. R. Soc.* B, 248, 135–204

LECLERG, E. L., W. H. LEONARD and A. G. CLARK (1966) *Field plot techniques*

MACKAY, J. R. (1958) 'The Anderson River map-area, N.W.T.', *Mem. geogr. Brch Can.* 5

MACKAY, J. R. (1963) 'The Mackenzie Delta area, N.W.T.', *Mem. geogr. Brch Can.* 8

MALAURIE, J. N. (1952) 'Sur l'asymétrie des versants dans l'île de Diskö, Groenland', *C.r. hebd. Séanc. Acad. Sci., Paris* 234, 1461–2

MELTON, M. A. (1960) 'Intravalley variations in slope angles related to microclimate and erosional environment', *Bull. geol. Soc. Am.* 71, 133–44

PEARSON, E. S. and H. O. HARTLEY (1951) 'Charts of the power function for analysis of variance tests, derived from the non-central F-distribution', *Biometrika* 38, 112–30

SCHUMM, S. A. (1956) 'The role of creep and rainwash on the retreat of badland slopes', *Am. J. Sci.* 254, 693–706

SCHUMM, S. A. (1964) 'Seasonal variations of erosional rates and processes on hillslopes in western Colorado', *Z. Geomorph., Suppl.* 5, 215–38

SELLERS, W. D. (1965) *Physical climatology*

SMITH, H. T. U. (1949) 'Physical effects of Pleistocene climatic changes in non-glaciated areas', *Bull. geol. Soc. Am.* 60, 1485–516

STRAHLER, A. N. (1950) 'Equilibrium theory of erosional slopes, approached by frequency distribution analysis', *Am. J. Sci.* 248, 673–96; 800–14

THORSTEINSSON, R. and E. T. TOZER (1962) 'Banks, Victoria and Stefanssen Islands, Arctic Archipelago', *Mem. geol. Surv. Brch Can.* 330

NOTES

1. This is a shortened and revised version of B. A. Kennedy and M. A. Melton (1967) 'Stream-valley asymmetry in an arctic-subarctic environment: conditions governing the geomorphic processes', Unpubl. Research Paper 42, Arctic Institute of North America.

2. These data were most kindly made available by Mr M. A. Church, Department of Geography, University of British Columbia.

RÉSUMÉ. *Sur l'asymétrie et les formes des versants dans une région de permafrost au Northwest Territories du Canada.* Les formes des versants et de l'asymétrie de quelques vallons au Northwest Territories du Canada sont décrites. Ces vallons se trouvent à l'est du delta de la rivière Mackenzie, creusés dans des argiles et des sables peu consolidés dont le pendage est vers le nord à moins d'un degré, environ; le gel du sous-sol est pérenne. Bien que l'étendue de la région étudiée soit restreinte, elle présente néanmoins une assez forte diversité en relief, climat et processus; elle est donc peu semblable à l'idéal de la « région périglaciaire » classique. L'inclinaison moyenne des versants augmente d'amont en aval de l'escarpement en même temps que le relief agrandit et le climat arctique se transforme en sub-arctique. L'asymétrie moyenne des pentes maximums démontre parallèlement une variation. En amont, ce sont des versants orientés vers le nord qui se trouvent les plus forts; en aval, au contraire, ceux qui sont tournés vers le sud ont les pentes les plus raides. En chaque endroit l'asymétrie est significative selon les épreuves statistiques. Une comparaison entre la mesure de l'asymétrie consecutive à l'érosion par les fleuves et celle qu'on peut attribuer uniquement à la variation en orientation des pentes, démontre une moindre valeur en premier cas, ce qu'on ne trouve pas dans de comparables enquêtes dirigées hors de la région de permafrost.

FIG. 1. La région du delta de la rivière Mackenzie

FIG. 2. Les Caribou Hills aux environs du Reindeer Station

FIG. 3. Quelques exemples de profiles en amont des vallons aux Caribou Hills

FIG. 4. Profil schématique en aval d'un vallon aux Caribou Hills

FIG. 5. Répartitition des échantillons aux Caribou Hills

FIG. 6. Les intéractions entre les effets de l'aspect et du relief: A, les valeurs moyennes de pente maximum; B, les valeurs moyennes de l'étendue des pentes maximums; C, les valeurs moyennes de pourcent de la couverture végétale aux pentes maximums

---

ZUSAMMENFASSUNG. *Talasymmetrie und Hangformen einer Zone einiges Gefrornise in den Nordwestgebieten, Canada.* Talasymmetrie und Hangformen werden in Bezug auf die grösseren Schwankungen des Klimas in einer kleinen Zone

Sedimentgesteine anliegend an das Mackenzie Flussdelta untergesucht. Die Studienzone ist geomorphologisch sehr verschieden und passt nicht zum klassischen ‚periglazialen' Modell. Asymmetrie der Höchsthangwinkeln wird statistisch analisiert und man hat gefunden, dass sie zwischen (a) Zonen mit einem härteren Klima und niedrigem verfügbaren Relief—wo Hänge, die nach Nordenliegen, viel steiler als sie, die nach Süden liegen, sind—und (b) die Zone mit einem milden Klima und tieferen Tälern, wo Hänge, die nach Süden liegen, die steileren sind, umgekehrt sind. Hänge, die direkt unter der Kontrolle von Flusserosion sind, zeigen wenigere Reaktion auf die Verschiedenheiten im Masse der Grundkorrasion als auf die Variationen der Exposition, eine Entdeckung, die mit den Resultaten der Studien von Zonen ohne einiges Gefrornis nicht übereinstimmt.

ABB. 1. Der Gegend des Mackenzie Deltas (nach J. R. Mackay, 1963, Abb. 71)

ABB. 2. Die Caribou Hills in der Nähe von Reindeer Station (vereinefacht vom Vordruck des Blattes 107B/11E, Serie 1:50 000, Abteilung der Landvermessungen und Kartographie, Ministerium der Bergwerke und technischen Vermessungen, Ottawa)

ABB. 3. Einige Beispiele der Querschnitte in den höheren Tälern, Caribou Hills

ABB. 4. Diagrammatisches Querprofil von einem Schnitte in einem niedrigeren Tale

ABB. 5. Der ganze Versuchsentwurf, Caribou Hills

ABB. 6. Wechselwirkung der Einflüsse von Exposition und topographische Lage: A, durchschnittlichen ‚Strahler' Winkeln: B, durchschnittliche Hänge der ‚Strahler' Winkelquerschnitte: C, durchschnittliche prozentige Pflanzendecke der ‚Strahler' Winkelquerschnitte

Modification of levee morphology by erosion in the Mackenzie River delta, Northwest Territories, Canada

DON GILL

(*Associate Professor of Geography, University of Alberta*)

Revised MS received 29 October 1971

ABSTRACT. While the unequal deposition of sediments controls the gross morphology of deltas, erosional processes also affect the distribution of sediments. In addition to normal fluvial erosion, levee morphology in an Arctic delta may be significantly modified by wave action, ice abrasion, slumping and solifluction, and thermal degradation of perennially frozen sediment. This paper analyses the action of these processes and shows how they affect levee retreat and rates of aggradation and progradation of slip-off slopes and point bars along shifting channels in the Mackenzie River delta.

FEW depositional features are fashioned as intricately as the extensive masses of alluvium that form the earth's deltas. The subaerial portion of a delta owes its morphology and very existence to irregular deposition of sediment, and most students of floodplain landforms have stressed the processes of alluviation which shape the deltaic surface (R. J. Russell, 1936; H. J. Harper, 1938; A. Sundborg, 1956; E. Kuiper, 1960; T. H. Schmudde, 1963; S. Dahlskog, 1966; V. Axelsson, 1967; and others). Few workers, however, have studied the counteractive force of erosion in modifying delta geomorphology, particularly in Arctic deltas. Russian permafrost experience has emphasized the process of erosion by thermo-erosional niche formation (A. I. Gusev, 1952; T. N. Kaplina, 1952; V. S. Lomachenkov, 1959) but other forms of deltaic erosion have not received equal attention, particularly in North America. The work of H. J. Walker *et al.* (1964, 1966, 1969) in Alaska's Colville River delta and in the Blow River delta of northern Yukon is an exception to the general tendency to ignore erosional processes, but these workers, too, emphasize thermal action. The purpose of this paper is to demonstrate the degree to which a wide range of erosional processes modifies levee morphology in an Arctic delta.

The Mackenzie is the longest river in Canada and one of the ten longest in the world. In North America, its drainage basin is exceeded in size only by that of the Mississippi. From the head of the Mackenzie's most remote tributary it is more than 4000 km to the point where it empties into the Beaufort Sea. That this great river periodically overflows its banks is manifested by the delta that it has created covering 6500 km².

The Mackenzie delta is of the estuarine type, being bounded on the west by the Richardson Mountains and on the east by the Caribou Hills. It is located immediately east of the border between Yukon Territory and the Northwest Territories. Figure 1 indicates its location as well as the area investigated during this study.

123

PROCEDURE

The following methods were used to measure the extent and rapidity of erosion during the 1966–67 and 1971 field seasons:

(1) Rates of cut-bank erosion were measured in 1967 by referring to stakes placed in selected positions during the previous field season. Rates of erosion were also measured by periodically surveying the position of tree stems relative to their decreasing distance from the edges of cut-banks; similar comparative measurements were made of turf overhangs.

(2) Wave erosion of slip-off slopes was measured after several periods of high winds to determine the degree to which rates of seasonal progradation and aggradation are retarded by this process. This was accomplished by comparing the known depths of sediment deposited during the 1966–67 and 1971 flood periods to the amount subsequently removed; erosion of this type creates miniature step-like scarps[1] and the amount removed can be approximated by projecting a line to re-create the previous surface.

(3) To measure the rate of scour-and-fill along a point bar, ten wooden dowels were driven into the unvegetated, prograding point-bar face, and a metal chain approximately 2 m in length was loosely attached to each. A ring and short section of each chain was placed on the surface (pointed down-channel) and the dowel marked at that level. The long section was buried to act as an anchor during flood. As scour occurs, the short length of chain becomes progressively lower until the base of the scour zone is reached, after which both the depth of erosion and the amount of sediment deposited after scouring takes place may be measured. The net increase or decrease in the profile relative to the previous year's surface is obtained by comparing the chain-dowel measurement with the amount of new sediment on the top of the chain.[2]

RESULTS

Three deltaic forms are significantly modified by erosion in the Mackenzie River delta. In order of descending degree of alteration they are: cut-bank levees, slip-off slopes, and point bars.

Cut-bank levees

Along shifting sections of distributaries[3] in the Mackenzie delta, lateral destruction of the floodplain by erosion is intermittently accompanied by alluviation along the crest of the concave bank near the channel edge. This form of levee is widespread throughout the northern (distal) half of the delta where levees are sufficiently low that overbank flow and accompanying sedimentation may occur simultaneously with erosional retreat of the cut-bank face. As indicated by J. R. Mackay (1963) the amount of sediment deposited along the backslope of the aggrading levee is in equilibrium with the rate of recession along the bank face so that this type of levee migrates laterally across the floodplain without greatly altering its cross-sectional dimension or form. Because of its distinct morphology and genesis it is categorized separately and is here termed a 'cut-bank levee'.

Alluvial deposits containing large fractions of silt and clay are cohesive and not easily eroded (L. Arnborg, 1948; Axelsson, 1967). Since alluvium in the Mackenzie delta is fine-grained (Fig. 2), high-cohesion slopes are maintained. Most cut-banks are thus nearly vertical, and where turf is present to bind the soil further, overhangs develop (Fig. 3). The presence of permafrost is also important in permitting vertical cut-banks to develop.

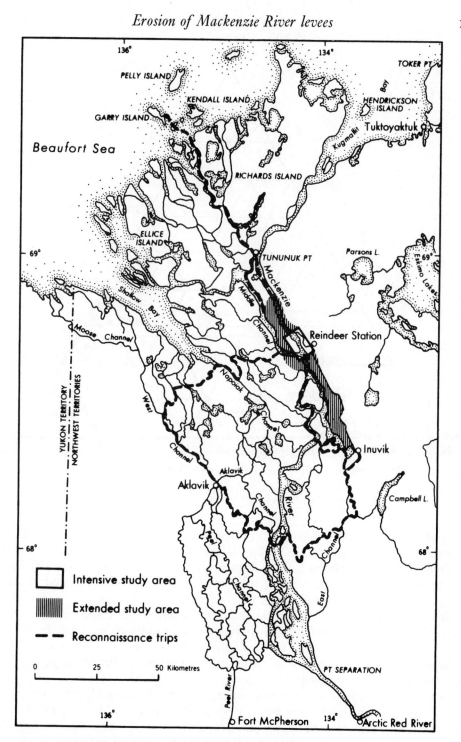

FIGURE 1. Study area within the Mackenzie River delta, Northwest Territories

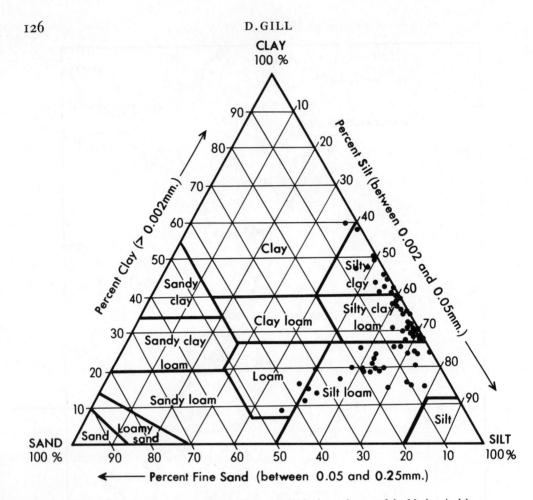

FIGURE 2. Classification of sixty-five sediment samples in the study area of the Mackenzie delta

Frozen sediment is resistant to erosion and slumping, and thus retrogression of a cut-bank can occur at a pace no greater than thawing of the bank face permits. As Walker and Arnborg (1966) also observed in the Colville delta, the thawed layer along a cut-bank forms on a plane parallel to the exposed surface; thus, once a vertical bank is initiated by undercutting, permafrost helps to perpetuate it.

While channel shifting is not pronounced in the Mackenzie delta, cut-banks retreat quite rapidly in certain localities. Such shifts are, however, caused primarily by processes other than normal fluvial erosion. Wave action, ice abrasion, and slumping[4] are of greater significance, and these processes in combination cause most channel migration. Four progressive steps are recognized:

(1) In late May and early June before flooding reaches a maximum height, the faces of many cut-banks, particularly those facing south, develop a thawed layer which may attain a depth of 0·5 m. Minor slumping and solifluction then begin, but the amount of material moved is relatively small. A thin veneer of thawed sediment may be removed by water when flooding begins, but this too is minor.

FIGURE 3. South-facing vertical cut-bank along an actively shifting distributary. Note overhanging turf and slumped spruce tree, indicators of rapid bank retreat. Water is at mid-flood-stage, 11 June 1967. Channel flow is from right to left

(2) Less bank erosion than might be supposed takes place during the flood period. At this time, because of the large amount of ice carried by channels (Fig. 4), water temperatures remain nearly constant just above freezing point (see the temperature trace for May 26–June 6 in Figure 5), and thus little water-induced thermal bank erosion can occur. Furthermore, rising floodwater effectively blocks solar radiation and prevents thawing by direct insolation. Although great quantities of ice are transported through the Mackenzie delta during the break-up of the ice, much of it is confined to the central section of the channel (coincident with that portion occupied by mean summer water level) and causes little damage to the banks. Similar observations were made by Walker and J. M. McCloy (1969) in the Blow and Colville River deltas. First to be transported through the distributaries is the delta ice itself. It normally 'candles'[5] during break-up (Fig. 4) and is not sufficiently solid to be an effective erosive tool. Candle-ice may actually form a protective shield against potential ice erosion. A band of candled ice often becomes lodged against a cut-bank where it remains stationary while other ice moves down-channel immediately against it. A distinct shearline develops which separates the moving from the non-moving ice (Fig. 4); under such conditions, bank erosion is minimized. Once Mackenzie River ice begins to pass through the delta, however, ice erosion does occur along certain banks. The river ice is usually much more massive and is less likely to be weakened by candling. As a result, it is an active erosional agent in some locations, especially where channels divide or along the outside bends of sharply inflected meanders (Fig. 6).

(3) After break-up when flood levels subside and ice is no longer present, channel temperatures rise rapidly (Fig. 5). Bank thawing now accelerates owing to a greater heat exchange with channel water and, with reduced water levels, the banks are no longer protected from insolation.[6] At this time, slumping of south-facing cut-banks may remove large blocks of levee material. Retreat along one south-facing cut-bank averaged 3 m in 1966 and 2·5 m in

1967. Between 1967 and 1971, retreat by slumping averaged 2·25 m in this location. Slumping was also measured in other locations: during 1966 one 0·5 km section of cut-bank retreated an average of 2 m. This bank was re-surveyed in the spring and in the fall of 1967, and it was found that during the three summer months it retreated approximately the same distance—again averaging nearly 2 m. In both years the deviation from these figures was small, indicating that recession was quite uniform. Figure 7 illustrates the amount of retreat which took place along the bank during this period.

FIGURE 4. Ice-choked distributary during flood stage. A band of candle-ice often lodges against a cut-bank during break-up and protects the shoreline from ice abrasion. Ice to left of dashed shear-line is stationary while that on right is moving forward at approximately 8 km/hour. 3 June 1967

(4) The fourth consecutive erosional process is that of erosion by wave-action (hereafter termed 'strandline erosion'). As distributary levels fall after flood, a sequence of elevated miniature shorelines is formed in the sediment of channel banks. Such features owe their existence to wave erosion, and in some portions of the delta (especially near the Caribou Hills where intense easterly gravity winds are common) strandline erosion is an important agent in the redistribution of cut-bank sediments.

(5) The final process is bank recession caused by thermal and fluvial erosion of permafrost at the water line (Fig. 8). Russian workers refer to such undercutting as 'thermo-erosional niche' formation (termoerozionaja niza; Gusev, 1952). According to Kaplina (1952) more undercutting is accomplished in such niches than along the exposed bank face because the exchange of heat between water and a solid at a given temperature is more intensive than between air and the same solid. Of equal importance, thawed material is rapidly removed from thermo-erosional niches by wave action which continues to expose the permafrost face to thermal erosion. Walker and McCloy (1969) report that in the Blow and Colville deltas, sediment from such niches is transported principally by fluvial erosion, but in the Mackenzie delta, wave action is considerably more important. This is probably because of the greater fetch of the much larger channels in the Mackenzie delta. Where long sections of north-west—south-east oriented distributaries occur, prevailing storm winds generate high-energy waves (wave length, height, and energy are increased when distributary current and wind are opposed). The period of greatest undercutting occurs in mid-July when water temperatures may reach as high as 19°C (Fig. 5). At this temperature, water is a most effective thermal sculptor

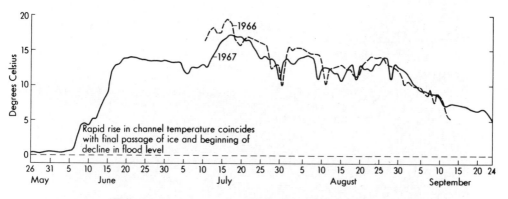

FIGURE 5. Daily water temperatures in a distributary of the Mackenzie delta, 1966–67

FIGURE 6. Massive river ice during break-up is an effective erosional tool when it is forced against the outside bend of sharply-inflected meanders. This levee is south-facing; thus, an easily eroded layer 65 cm deep had thawed by 8 June 1967, when this cake of ice 1·5 m thick was forced against it

and, when accompanied by high winds, a wedge-shaped niche may penetrate 3 to 5 m into the frozen sediment in less than 48 hours. At this time, large blocks of permafrost fall into the channel where they are removed within 1 to 2 weeks by thermal and fluvial erosion (Fig. 9). Along one such shifting channel (in the south-west corner of the intensive study area, Fig. 1), a comparison of its 1952 position (as determined from aerial photographs) with its surveyed location in 1971 indicates that, during a 19-year period, the channel shifted approximately 180 m, or nearly 10 m/year. It is significant that

FIGURE 7. Rapid cut-bank retreat; approximately 2 m of recession during the summer of 1967 exposed these
buried *Salix* stems. Direction of channel flow is toward the observer. 29 August 1967

other workers have measured comparable rates of bank recession in other Arctic deltas.
Walker and Arnborg (1966) surveyed 10 m of retreat in an area of the Colville delta during
1962. Lomachenkov (1959) also reports a similar rate of bank collapse by the formation of
thermo-erosional niches in a section of the Indigirka delta of eastern Siberia. These few
measurements suggest that thermo-erosional processes in Arctic deltas may be relatively
uniform through time.

Slip-off slope levees and point bars

Three types of erosion actively sculpture slip-off slopes in the Mackenzie delta—cross-
levee gullying, strandline erosion, and the 'cut' of normal scour-and-fill processes that
accompany the construction of point bars and slip-off slopes.

Cross-levee gullying Behind most point bars and slip-off slopes, a meander-scroll
depression exists which may fill with floodwater and/or meltwater (D. Gill, 1971b).
During the lowering of channel levels after flood-stage, this water may leave the depression
to erode small gullies on its way to the receding distributary. Although this form of erosion
is not widespread (since most meander-scroll depressions are of a sufficient height to be
above flood level and most have high rims which act as dams), gullying nevertheless alters
the morphology of many slip-off slopes (Fig. 10).

Strandline erosion This process is a most important modifier of virtually every prograding slip-off slope in the Mackenzie delta, and it also affects point bars (Fig. 13).
During flood recession, strandline erosion begins to rework the sediment previously

deposited; Figure 11 shows several 30–70 cm scarps eroded into new alluvium by wave action. As the post-flood season advances, slip-off slopes continue to be sculptured by strandline erosion, with the result that a series of steps is formed. Viewed aerially (Fig. 12), it is readily apparent that wave erosion is important in altering the morphology of slip-off slopes. It was calculated by comparing measurements of wave-cut scarps to known depths in 1966–67 and 1971 alluvial deposits that 40 per cent of the potential progradation of the slip-off slope levee shown in Figure 12 was retarded by strandline erosion during these 3 years. The rate of progradation of slip-off slopes is most affected, since erosion occurs primarily along their leading edge; little or no strandline erosion can occur within the heavily vegetated portion owing to the cushioning effect which dense willow stems have on waves and fluvial action (Gill, 1971a). Note in Figure 11 that the most inland strandlines are just within the *Salix-Equisetum* association. Aggradation of slip-off slopes thus continues apace; this is one reason why such deposits in the Mackenzie delta form a corrugated surface—the leading edge of a slip-off slope aggrades more rapidly than it is permitted to prograde, with the result that the smoothing influence of alluviation is reduced.

FIGURE 8. Thermo-erosional niche 5 m deep developed during July 1967. Water is level with central portion of niche. Note ice-wedge and slumped block on left. Scale is provided by *Salix* and *Alnus* stems on upper left which average 6 cm in diameter

Scour-and-fill As indicated by L. B. Leopold *et al.* (1964), scour may take place at nearly all sections along a reach of a river during flood, and the process of scour-and-fill is one of the mechanisms governing the adjustment of the size and shape of channels. Figure 13 (p. 135) illustrates the amount of erosion and subsequent deposition that took place during the 1968 flood period along two transects measured transversely to a typical point bar. Scour initially removed up to 32 cm of sediment which was subsequently replaced. The nearly balanced cut-and-fill (there is a net gain of several cm of sediment) indicates that the rate of progradation at this point bar was relatively slow. This information is based on only one year's readings, however, and may not represent the true degree of equilibrium maintained by this channel section. It is presented to demonstrate that, while a point bar is primarily a feature of alluvial accretion, erosion during flood stage can be an important associated process.

FIGURE 9. Rapid bank retreat resulting from thermo-erosional niche formation. Note large slumped block in left centre. Extreme bank recession occurred in this location between 1966 and 1971. Channel flow is toward the camera. 20 July 1967

FIGURE 10. Cross-slope gully indicating that the morphology of a prograding slip-off slope may be considerably altered by fluvial erosion. Also note strandline erosion

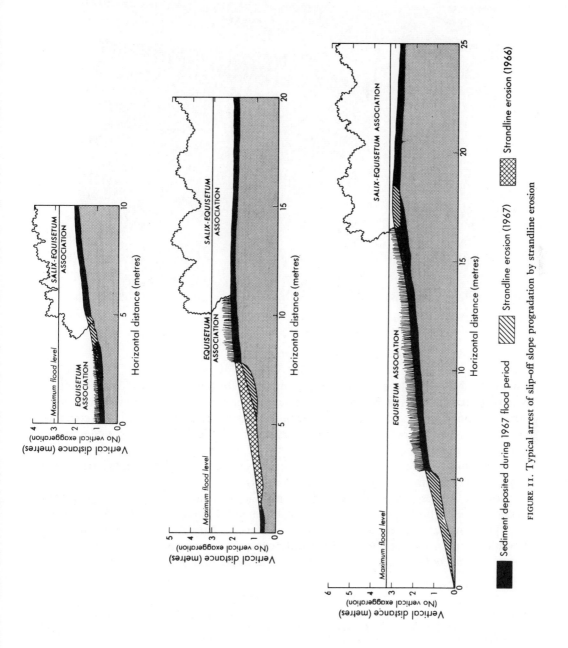

FIGURE II. Typical arrest of slip-off slope progradation by strandline erosion

FIGURE 12. Aerial view of a strandline scarp along a slip-off slope on 30 August 1967, illustrating the large
amount of sediment which was removed that summer by wave action

SUMMARY AND CONCLUSIONS

It has been shown that levee morphology in an Arctic delta may be considerably modified
by erosion. Ice abrasion, slumping and solifluction, and normal fluvial erosion shape levee
forms to a moderate extent, while wave action independently, and wave action in conjunc-
tion with thermal degradation of permanently frozen sediments, is very significant in
altering levee morphology. Although extremely rapid bank retreat through the formation of
thermo-erosional niches is not widespread in the Mackenzie delta, channel shifting is
greatly accelerated in some locations by this process.

Most of the destruction of cut-bank levees is accompanied by deposition along their
backslopes; thus, this levee form is seldom totally removed by erosion. Only where thermo-
erosional niches are active is the floodplain destroyed laterally without accompanying
alluviation.[7] In the greater portion of the study area, the morphology of cut-bank levees is
thus in dynamic equilibrium between the opposing forces of erosion and deposition.

ACKNOWLEDGEMENTS

Most of the data on which this work was based were collected in 1966–68 as part of a study for a Ph.D. at the Univer-
sity of British Columbia (Gill, 1971c). The author thanks A. L. Farley, V. J. Krajina, J. Ross Mackay, J. C. Ritchie,
O. Slaymaker and J. K. Stager for critical comments. The study was supported by funds provided through J. Ross
Mackay from the National Research Council of Canada, Department of Indian Affairs and Northern Development,
U.B.C. Research Funds, and the Department of Energy, Mines, and Resources. A University of Alberta Research
Grant (55-32254) to the author permitted him to follow-up the work during the 1971 field season. Peter J. Lewis,
University of British Columbia, and Steve Gill helped in the field for which the author is grateful. Special apprecia-
tion is expressed to Ross Mackay for his generous support. The University of Alberta kindly contributed towards
the cost of illustrations.

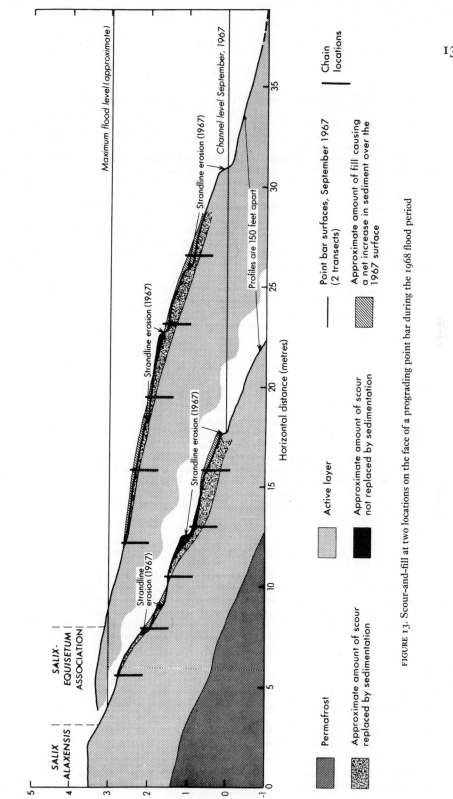

FIGURE 13. Scour-and-fill at two locations on the face of a prograding point bar during the 1968 flood period

135

NOTES

1. The 'wave-cut scarp' of V. Axelsson (1967, p. 116).

2. After J. Ross Mackay, University of British Columbia (personal communication).

3. Channels in the Mackenzie delta do not meander in the classic sense; they shift instead (Mackay, 1963).

4. Bank undercutting and associated true rotational slumping are rare in the Mackenzie delta. Removal of cut-bank material is a complex multiple process, combining true slumping with soil flow (or soil creep). For simplification, this process is referred to as slumping.

5. 'Candle-ice' refers to the long, vertical ice crystals which form perpendicular to the freezing plane during the freezing of natural water bodies. At break-up, preferential melting along the crystal interstices separates the ice body into a mass of individual 'candles' that, in the Mackenzie delta, normally average somewhat less than 1 m in length and 5 cm in diameter (Fig. 4).

6. Thawing of south-facing cut-banks at this time is accelerated by the sun's reflection from the water surface (Mackay, personal communication).

7. Long-term erosion of levees may exert a strong influence on the ecological succession of Mackenzie delta vegetation. The intermittent destruction of levees is partially responsible for maintaining white spruce (*Picea glauca*) as the dominant species of the climax ecosystem. If a section of relatively unflooded upper levee surface remained intact for a sufficient period, autogenic processes would result in an increasingly thick organic mull/mor cover, acidification of the soil, lowered soil temperatures, and a decrease in the thickness of the active layer. Under such conditions, white spruce would be gradually replaced by acidic- and moisture-tolerant black spruce (*Picea mariana*) which is now adjacent to (but does not occur in) the delta. The time necessary to establish such conditions is probably greater than the present age of most segments of surface owing to the continued replacement of mature levees by *terra nova* through distributary shifting. The process of erosion may thus be interrupting a more extended successional sequence than is now measureable in the Mackenzie delta.

REFERENCES

ARNBORG, L. (1948) 'The delta of Ångermanälven', *Geogr. Annlr* 30, 673–90

AXELSSON, V. (1967) 'The Laitaure delta, a study of deltaic morphology and processes', *Geogr. Annlr* 49-A, 1–127

DAHLSKOG, S. (1966) 'Sedimentation and vegetation in a Lapland mountain delta', *Geogr. Annlr* 48-A, 86–101

GILL, D. (1971a) 'Damming the Mackenzie: a theoretical assessment of the long-term influences of river impoundment on the environment of the Mackenzie River delta, Northwest Territories', *Proc. Peace-Athabasca Delta Symp.*, Univ. of Alberta (14–15 January 1971), 204–22

GILL, D. (1971b) 'From helicoidal flow to the beaver stretcher: eight steps in an ecologic linkage', *Preconf. Publ. Pap.* (Annual Meeting of the Canadian Association of Geographers, Univ. of Waterloo), 9–15

GILL, D. (1971c) 'Vegetation and environment in the Mackenzie River delta, Northwest Territories: a study in subarctic ecology', Unpubl. Ph.D. thesis, Dept. of Geography, University of British Columbia.

GUSEV, A. I. (1952) 'On the methods of surveying the banks at the mouths of rivers of the polar basin', *Trans. Inst. Geol. Arctic* (Leningrad), 107, 127–32

HARPER, H. J. (1938) 'Effect of silting on tree development in the flood plain of Deep Fork of the North Canadian River in Creek Co., Oklahoma', *Proc. Okla. Acad. Sci.* 18, 46–9

KAPLINA, T. N. (1952) 'Some features of outwash of shores composed of permanently frozen rocks', *Trans. Oceanogr. Commn Acad. Sci. U.S.S.R.* 4, 114–15

KUIPER, E. (1960) 'Sediment transport and delta formation', *J.Hydraul. Div. Am. Soc. civ. Engrs* 86, 55–68

LEOPOLD, L. B., M. G. WOLMAN, and J. P. MILLER (1964) *Fluvial processes in geomorphology*

LOMACHENKOV, V. S. (1959) 'The preparation of maps on the morphology and dynamics of the banks of low river terraces in an area of permafrost', *Trans. Inst. Geol. Arctic* (Leningrad), 65, 157–9

MACKAY, J. R. (1963) 'The Mackenzie delta area, N.W.T.', *Mem. geogr. Brch Can.* 8, 202 pp.

RUSSELL, R. J. (1936) 'Physiography of the lower Mississippi delta', Louisiana Consn. Dept. *Bull.* No. 8, 3–199

SCHMUDDE, T. H. (1963) 'Some aspects of land forms of the lower Missouri River floodplain', *Ann. Ass. Am. Geogr.* 53, 60–73

SUNDBORG, A. (1956) 'The River Klarälven, a study of fluvial processes', *Geogr. Annlr* 38, 280–91

WALKER, H. J., and J. M. MCCLOY (1969) 'Morphologic change in two Arctic deltas', *Arctic Inst. N. Am.*, Final Rep. of Grant no. DA-ARO-D-31-124-G832, 91 pp.

WALKER, H. J. and L. ARNBORG (1964) 'Morphology of sand dunes in an arctic delta', *Abstr. Pap. 20th int. geogr. Congr.* (London), 116

WALKER, H. J., and L. ARNBORG (1966) 'Permafrost and ice-wedge effect on riverbank erosion', *Proc. Permafrost int. Conf.* (11–15 November 1963, Lafayette, Ind.), 1287, 164–71

RÉSUMÉ. *Modification de la morphologie des levées provoquée par l'érosion dans le delta du Mackenzie, Territoires du Nord-ouest.* Tandis que le dépôt irrégulier de sédiments détermine la morphologie générale des deltas, des mécanismes d'érosion affectent aussi la répartition des sédiments. À l'érosion fluviale normale, il faut ajouter l'action des vagues, l'abrasion de la glace, les éboulements, la solifluction, et la dégradation thermique de sédiments gelés en permanence, phénomènes qui peuvent modifier d'une manière importante la morphologie des levées d'un delta dans l'Arctique. Cette thèse se propose d'analyser la manière dont ces mécanismes affectent le recul des levées, et le degré d'alluvionnement et d'avancée du talus des rives convexes et le long des bras divageants du delta du Mackenzie.

FIG. 1. Aire d'études dans le delta du Mackenzie, Territoires du Nord-Ouest

FIG. 2. Classification de soixante-cinq échantillons de sédiments dans l'aire d'étude du delta

FIG. 3. Rive concave verticale à regard sud le long d'un bras divageant actif. Remarquez la couche de tourbe en surplomb et le *Picea* ayant glissé, tous deux indicateurs du retrait rapide de la berge. L'eau est à moitié de son maximum aux hautes eaux. 11 juin, 1967. L'écoulement se fait de droite à gauche

FIG. 4. Bras couvert de blocs de glace au moment des hautes eaux. Une épaisseur de chandelles de glace se loge souvent contre la rive concave au moment de la débâcle et la protège de l'abrasion de la glace. À la gauche de la ligne de dislocation (marquée par des tirés) la glace est stationnaire, tandis qu'à droite elle est mouvante. Sa vitesse de déplacement est approximativement de 8 km par heure. 3 juin, 1967

FIG. 5. Températures quotidiennes de l'eau dans un bras du delta, 1966–67

FIG. 6. Au moment de la débâcle, les énormes blocs de glace poussés contre la rive concave des méandres sont un agent efficace d'érosion. Cette levée est exposée au sud; de sorte qu'une couche de 65 cm, qui avait dégelé le 8 juin 1967, était facilement érodé par ce bloc de glace de 1,5 m d'epaisseur

FIG. 7. Retrait rapide de la rive concave; approximativement 2 m de recul durant l'été de 1967 a découvert ces troncs de *Salix*. L'écoulement est face à l'observateur. 29 août, 1967

FIG. 8. Une niche thermo-érosive profondement développée durant le mois de juillet 1967. L'eau est au niveau de la partie centrale de la niche. Remarquer le coin de glace et le bloc effondré à gauche. L'échelle est donnée par les troncs de *Salix* et *Alnus* en haut à gauche qui ont en moyenne 6 cm de diamétre

FIG. 9. Le recul rapide de la berge résulte de la formation de la niche thermo-érosive. Remarquer les grands blocs effondrés au centre gauche. Un recul important de la berge s'est produit à cet endroit entre 1966 et 1971. L'écoulement est face à l'observateur. 20 juillet, 1967

FIG. 10. Le ruissellement transversal indique que la morphologie d'une rive convexe en construction peut être considérablement altérée par l'érosion fluviale. Remarquez également l'effet de l'érosion par les vagues

FIG. 11. Arrêt typique de l'avancée de la rive convexe par l'érosion par les vagues

FIG. 12. Photographie aerienne d'un escarpement provoqué par les vagues le long de la rive convexe le 30 août, 1967, qui indique la grande épaisseur de sédiments enlevés cet été par l'action des vagues

FIG. 13. Érosion et depôts à deux endroits sur le versant de la rive convexe au moment des hautes eaux de 1968

ZUSAMMENFASSUNG. *Modifikation der Uferdamm-Morphologie durch Erosion im Mackenzie Fluss-Delta, Nordwestgebieten.* Während die ungleiche Sedimentbildung die gesamte Morphologie eines Deltas bestimmt, können Erosionsvorgänge auch die Verteilung der Sedimente beeinflussen. Zusätzlich zur gewöhnlichen Flusserosion wird ein arktisches Delta auch bedeutend modifiziert durch Wellenbewegung, Eisabschürfung, Sedimentensenkung und Solifluction, so wie auch die thermische Herabsetzung des beständig verfrorenen Sediments. Diese Abhandlung untersucht, wie diese Vorgänge, das Zurückweichen des Uferdamms und der Grad der Anschwemmung und Vorrückung der Gleithängen und Sedimentbänken entlang den wechselnden Flusskanalen im Mackenzie Fluss-Delta, beeinflussen.

ABB. 1. Untersuchungs Gebiet das Mackenzie Fluss-Delta, Nordwestgebieten

ABB. 2. Die Klassifizierung fünf-und-sechzig Sedimentproben aus dem Untersuchungs Gebiet das Mackenzie Fluss-Delta

ABB. 3. Ein Südliegender senkrechter geschnittener Abhang entlang einem activen verschiebenden Flussarm. Beachte den überhängenden Torf und die gesunkene Fichte, ein Zeichen von Schnellen Rückzug des Abhanges. Der Fluss ist im Mitte dem Hochwasser Stadium 11 Juni, 1967. Flussströmung von Rechts nach Links

ABB. 4. Eis verstopften Flussarm während der Flut. Ein streifen Kerzen eis lagert sich oft gegen den Abhang während des Aufbruches und beschützt das Ufer von Eisabschürfungen. Eis links von der Unterbrochene Scherlinie ist stationär während das auf der rechten Seite sich ungefähr 8 km/Stunde vorwärts bewegt. 3 Juni, 1967

ABB. 5. Tägliche Wasser Temperaturen in ein Flussarm der Mackenzie Delta, 1966–67

ABB. 6. Massives Flusseis während des Aufbruches ist ein wirksames Schleifwerkzeug wenn es gegen die äussere Kurve einen Stark Mäander gedrückt wird. Dieser Uferdamm ist Südliegend, daher eine leicht Zerfressbare Schicht 65 cm dick wurde bis 8 Juni, 1967 aufgetaut als diese 1,5 m dicke Eisklotz gegengepresst wurde

ABB. 7. Ein schneller abhand Rückzug: ungefähr 2 m zurückgehen des Abhanges während des Sommers 1967 auf deckte diese begrabene *Salix* Stengel. Der Flussarm fliesst auf den Beobachter zu. 29 August, 1967

ABB. 8. Eine Thermozerfressene Nische 5 m tief entwickelt im Juli 1967. Das Wasser ist eben mit dem zentral Teil der Nische. Beachte den Eiskiel und gesunkenen Klotz im Centrum der Nische. Das Grösserverhältnis ist durch die *Salix* und *Alnus* Stengel oben links gezeigt, die ein durchschnitts Durchmesser von 6 cm haben

ABB. 9. Ein Schneller Abhang Rückzug bewinkt durch die thermoerosion Nische Formation. Beachte den grossen gesunkenen Klotz links Mitte. Extreme Abhang rückgang ist im diesem Orte zwischen die Jahre 1966–71 vorgekommen. Der Flusskanal fliesst auf der Kamera zu. 20 Juli, 1967

ABB. 10. Querhang Rinne, ein Hinweis dass die Morphologie eines Vorrückenden Gleithang sich sehr verändern kann durch auswashungen. Beachte auch die Erosion der Auschwemmungslinie

ABB. 11. Typischer arrest die Gleithang Vorrückung durch Erosion der Auschwemmungslinie

ABB. 12. Luftaufnahme eine auschwemmungslinie Eskarpe entlang ein Gleithang, 30, August 1967, erklärt das viele Sediment, dass im Sommer durch Wellenbewegungen weggenommen wurde

ABB. 13. Abscheuerungen und Fülle an zwei Stellen an der Fläche eines Vorrückenden Sedimentbankes während der 1968 hochwasser Periode

Relationships between process and geometrical form on High Arctic debris slopes, south-west Devon Island, Canada

P. J. HOWARTH
Assistant Professor of Geography, McMaster University

AND J. G. BONES
Graduate Student in Geography, University of Alberta

Revised MS received 29 October 1971

ABSTRACT. Measurements of debris slope profiles have been made at four different locations on south-west Devon Island, N.W.T., Canada. At these locations, the nature and effects of basal erosion, rockfall and of processes dominated by meltwater have been described. Analysis of the slope angles indicates that there are significant differences in angle between slopes affected by different processes. The steepest slopes, tending to be convex in form, are produced by basal erosion which results from both ice push and storm waves. In non-coastal locations, it was found that slopes dominated by rockfall are significantly steeper than those affected by processes involving the presence of water, although both types of slope tend to be concave in form. The effects of these two processes are also seen in the significant differences between talus sheets (rockfall dominant) and talus cones (meltwater dominant). In coastal areas, however, rockfall slopes and avalanche slopes are both steep because debris is rafted away by sea ice to prevent basal accumulation. There appears to be little difference in geometrical form between High Arctic and mid-latitude slopes if angles and profiles are compared on slopes where similar processes operate in the two environments.

WITH the exception of A. Rapp's (1960) report on Spitsbergen, the active periglacial environment of the High Arctic has been neglected as a location for studies of debris slopes. Consequently, it is not known if the relationships between process and morphology which occur on mid-latitude and high alpine debris slopes are similar to those of the High Arctic environment. A series of studies has been carried out on selected debris slopes of south-west Devon Island, N.W.T., Canada, to clarify the relationships which may exist between process and morphology (J. G. Bones, 1971). Morphology is taken to comprise both the geometrical and the sedimentary form of the slope. This paper, however, considers only the geometrical form (profile angle and form) and the way in which it is influenced by different slope processes. A discussion of process and sedimentary characteristics is to be published elsewhere.

AREA OF STUDY

The study area in the south-west of Devon Island (74°30′N and 91°10′W) is bordered by three major water bodies; Gascoyne Inlet to the west, Lancaster Sound to the south and Radstock Bay to the east (Fig. 1). The western part of the island is underlain by Palaeozoic strata, the rocks in the Radstock Bay area being primarily Upper Silurian in age (Y. O. Fortier in Fortier et al., 1963; R. Thorsteinsson, 1958). They are of two main types. First, there are silty, argillaceous and dolomitic limestones which weather into comparatively

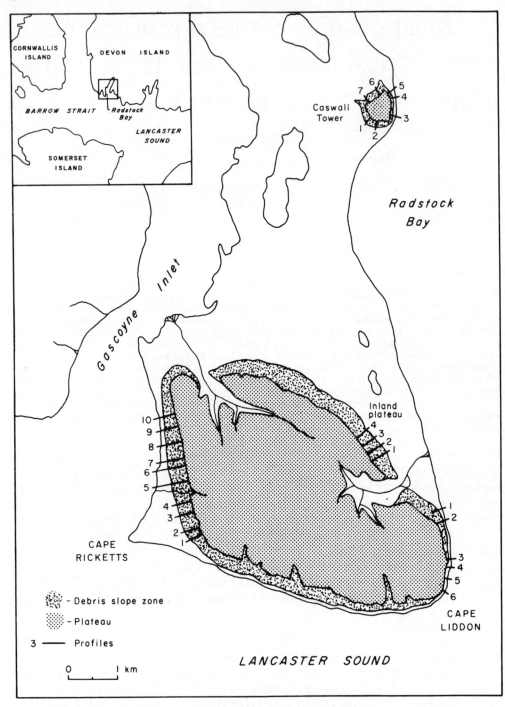

FIGURE I. Location of slopes investigated in the Radstock Bay area. Beach deposits occupy most of the remaining land surface

thin plates; and secondly, there are crinoidal limestones which are more blocky in appearance. With the very gentle dip of the strata towards the west and south, a basically uniform lithology exists on each slope. This permits the differences in slope form and development to be analysed as a product of the types and intensities of processes that occur.

The dominant geomorphological feature of the area is the 320–400 m high Barrow Surface, the erosional remnant of an older and higher surface thought to have been eroded subaerially during a Tertiary uplift (J. B. Bird, 1959). Within the study area, two isolated remnants of the plateau occur (Fig. 1), and it is the debris slopes developed beneath the edges of these remnants that have been investigated. In places, scree slopes begin right at the edge of the plateau; in the majority of cases, however, talus sheets and talus cones have developed beneath a steep cliff face. Some of the debris slopes extend down to the present active beach zone, while others end on the raised beaches of the coastal lowlands, which reach altitudes of approximately 100 m (Bird, 1967, p. 130). Consequently, debris slopes with different input and output environments were readily available for study.

Within the general setting, four specific localities were investigated (Fig. 1).

Caswall Tower

This plateau remnant, with an altitude of 196 m, is located 6·4 km along the shore from the western entrance of Radstock Bay. Extending north and west from the Tower are two elevated gravel beach-ridges, between which a well-developed talus sheet occurs. This grades northwards into coalescent talus cones. Moving eastwards, the type of slope alternates between talus sheets and talus cones developed below a well-dissected and increasingly steepening cliff face. The eastern side of the Tower borders on Radstock Bay and at high tide the slopes are separated from the sea by only a few metres of beach. The beach zone widens towards the south where the base of the slope is resting on raised beach-ridges. Seven slopes at Caswall Tower were studied.

Inland Plateau

Here scree slopes and occasional talus cones extend down from the plateau top into a wide depression behind a raised beach. Solifluction debris from the base of the slopes covers much of the surface of the depression. A wide band of resistant dolomitic limestone is visible about half way up the debris-covered rock face, and the occasional gullies developed in this rock face have resulted in the formation of several large talus cones beneath it. Two talus cones and two scree sheets were studied in this area.

Cape Liddon

The slopes considered lie along the south-eastern extremities of the Cape, below a well-dissected cliff face 200 to 300 m in height. Some slopes are developed over small rock platforms, the seaward edges of which often protrude from beneath the talus (Fig. 2). At high tide only a few metres of gravel separate the foot of the slopes from the sea. The Cape Liddon slopes face south and south-east in the direction of longest wave fetch, a situation which at times exposes them to basal erosion. Six debris slopes were analysed here.

Cape Ricketts

Slopes in this area are both talus sheets and cones which together extend continuously from south to north along the plateau edge. The slopes are protected from marine action by

FIGURE 2. A rock platform at Cape Liddon exposed by basal erosion

a series of well-defined marine terraces, a solifluction terrace and a wide expanse of raised beaches behind the modern beach zone. The width of this protective foreland decreases towards Gascoyne Inlet in the north. Plateau height and the amount of exposed free-face similarly decrease from south to north, and the northerly slopes appear to have over-ridden marine and solifluction terraces that once protected them. Ten Cape Ricketts slopes were studied.

METHODS

Tacheometric methods with a Wild T2 theodolite and survey staff were used to obtain profiles of the debris slopes. For each slope, two lines were surveyed from base to top to obtain a representative profile for both the left and right sides of the slope. On debris sheets, profile lines were surveyed parallel to each other, while on cones the lines converged towards the top. In this way, the profiles followed the lines of steepest slope. Survey points were chosen on the slope in order to portray the major variations in form that occurred.

The survey data were used to reconstruct the slope form as a series of straight-line segments. Profile analysis was carried out according to a method suggested by R. F. Stock (1968) which provides description, comparison and rank of slope forms in terms of the degree of conformity to a straight-line profile. The degree of deviation from this form is considered to reflect differential rates of a process or the presence of differing processes. Slope 'components' (convex or concave) are delimited and used to calculate indices of convexity or concavity for the whole profile. Slope lengths are standardized to allow the creation of dimensionless numbers for easier comparison between slopes. From these indices a sum and a ratio of indices are derived, which are used to classify the profile as one of six types:

1. straight
2. concave
3. concave with minor convexity
4. convexo-concave (equal)
5. convex with minor concavity
6. convex

In this paper, the method has been used to provide an objective classification for describing the slope forms. A. Young's (1964) approach to the study of profile form could not be attempted as it required both straight and curved slope sections of which only the curved may be convex or concave.

The limitations of slope profiles in this study must be realized, since the most that two representative profiles can offer is a very small impression of the overall slope form. Slope form varies over both time and space, and one must be wary, when using straight-line profiles, of extending the interpretation to events occurring over the whole surface of the slope (A. Jahn, 1968). Most of the results derived from the slope profiles of south-west Devon Island, however, were verified by field observations. As field studies covered spring, summer and early winter periods, it was possible to observe the major slope-forming and slope-modifying processes in operation. In the following three sections, the geometrical form of the slope profiles is considered in areas where different processes were observed.

GEOMETRICAL FORM AND BASAL EROSION

Along the coastline of south-west Devon Island, basal erosion can occur both by ice-push and by storm waves. For 10 months of the year this coastline is protected from basal erosion of any type by a complete cover of sea ice and beach ice (S. B. McCann and E. H. Owens, 1969; Owens and McCann, 1970). If strong on-shore winds occur during the break-up period, however, large ice floes can be pushed against coastal debris slopes. After the sea ice has disappeared, storm waves can also produce erosion. The Cape Liddon slopes and the seaward-facing slopes of Caswall Tower face south-east towards the direction of longest wave fetch and have comparatively deep off-shore zones. If marine and weather conditions are favourable, the debris slopes in these two areas can be attacked by basal erosion.

Two major storms occurred during the summer of 1969. The first took place during the period 25–26 July under conditions of partially-open water. Pack ice was pushed against the base of several Cape Liddon slopes by 22 m.p.h. (1·0 m/s) winds blowing from the south-east. Although the ice floes were forced against the bases of the slopes, the presence of permafrost minimized the amount of surface debris removed. Slabs of ice on some Cape Liddon slopes reached a maximum height of 20 m above beach level. The release of pack-ice pressure and the occurrence of high tides eventually led to the removal of most of the ice from the slopes.

The second storm took place during 11–13 August while Radstock Bay was ice-free. South-easterly winds gusting to 42 m.p.h. (1·9 m/s) generated storm waves with heights sufficient to overrun the beach zone at both Cape Liddon and Caswall Tower with the result that several talus slopes were eroded at base. Where most effective, wave erosion produced near-vertical cuts of 2 to 3 m in the slope base. On several Cape Liddon slopes, this erosion revealed the frontal sections of a rock platform, probably formed when sea level was relatively higher than at present (Fig. 2). Where erosion was less effective, basal areas of slopes were initially bared of surface debris, but not appreciably altered in form.

FIGURE 3. Ice and permafrost exposed at the base of a debris slope

FIGURE 4. Down-slope movement of material has buried the rock platform seen in Figure 2

The basal erosion, both by ice push and wave action, frequently revealed a layer of ice protruding from beneath the up-slope surface debris (Fig. 3). This layer is quite distinct from the permafrost table because of its coarse crystalline nature, its thickness and lateral variation, and the absence of debris locked within it. Such layers were observed in several different locations. They are interpreted as remnants of snow banks accumulated during the winter, which are subsequently compacted and recrystallized beneath debris brought down by rockfalls and snowmelt. The occurrence of deep snowbanks at the base of slopes was observed in October of 1969, and early in the summer of 1970 many such debris-covered areas were at first mistaken for actual extensions of the slope foot.

Following the storm of 11–13 August, readjustment of slopes to the effects of basal erosion was observed and recorded by photography at regular intervals for three sites along Cape Liddon. The sequence of events was similar at all sites, the only variation being in the rate and extent of readjustment. The initial basal cut exposed either the compacted snowbanks or the permafrost. Melting of this ice caused slope instability, and debris from immediately above the cut moved down-slope, thereby exposing permafrost higher up the slope. This is because the permafrost table slopes parallel to the surface profile. In this way the effect of ice melting gradually extended up-slope. Higher up the slope the larger fragments became affected and were removed to the base. When removed, the rate of exposure of permafrost decreased until a comparatively stable situation was achieved. The material deposited at the base of the slope eventually filled the erosion scar and covered the exposed rock platform (Fig. 4). A new slope foot was thus created with an angle similar to the one present before erosion.

A return to the sites early the following summer indicated that the phase of readjustment does not normally extend to the top of the slope in one season; rather, it reaches some critical height at which sufficient material has accumulated down-slope to enable slope instability to be minimized. Furthermore, some of the basal-cut areas revisited at Caswall Tower had not reformed over the winter. This indicates that readjustment to profile change may be affected by a lower magnitude of basal erosion and a lower rate of accumulation of rockfall material.

The five profiles measured on slopes affected by basal erosion fall into five of the different categories suggested by Stock (1968). The only form absent is 'concavity'. The three profiles from the most severely affected area, Cape Liddon, consisted of two dominantly convex slopes and one straight slope. This latter slope was comparatively steep and short, and the basal cut had already been eliminated before the slope was surveyed. At Caswall Tower, a less severe basal attack was observed and readjustment affected only the lower parts of the slopes. This proved insufficient greatly to alter the concave profile inherited from rockfall accumulation, so only minor convexities were introduced into the profile by basal erosion.

Long-term effects are seen in the overall slope angles (Table I), those for Cape Liddon (mean angle 38·8°) being greater than those for the seaward-facing slopes of Caswall Tower (mean angle 36·9°).

TABLE I

Classification of debris slope profiles ranked in order of increasing concavity for each site

Profile No.*		Dominant† process	Mean angle	Form¶
CAS6	(2)	M	33·60	4
CAS5	(1)	M	35·83	3
CAS3	(2)	B	37·02	4
CAS4	(2)	R	34·23	2
CAS1	(2)	R	32·10	5
CAS2	(1)	R	33·32	3
CAS2	(2)	R	33·94	3
CAS3	(1)	B	36·72	3
CAS1	(1)	R	33·02	5
CAS6	(1)	M	34·43	2
CAS5	(2)	M	33·40	3
CAS7	(1)	M	30·65	3
CAS4	(1)	R	32·73	2
CAS7	(2)	M	31·68	2
CL1	(1)	R	36·23	1
CL2	(1)	R	38·22	1
CL4	(1)	R	34·00	1
CL5	(1)	M	38·30	1
CL1	(2)	R	37·02	3
CL3	(1)	R	36·12	2
CL6	(1)	B	39·97	3
CL4	(2)	R	37·00	6
CL3	(2)	R	35·42	2
CL2	(2)	B	37·00	5
CL5	(2)	M	36·93	2
CL6	(2)	B	39·33	3

TABLE I—*contd.*

Profile No.*		Dominant† process	Mean angle	Form¶
IN1	(1)	R	33·50	3
IN1	(2)	R	33·22	2
IN4	(1)	R	31·75	2
IN4	(2)	R	31·68	2
IN3	(1)	M	22·53	2
IN3	(2)	M	21·92	2
IN2	(1)	M	19·78	2
IN2	(2)	M	19·75	2
RK2	(2)	R	33·40	3
RK4	(2)	R	33·65	3
RK4	(1)	R	33·60	2
RK3	(2)	M	26·85	3
RK10	(2)	R	34·12	3
RK2	(1)	R	32·90	2
RK7	(1)	M	25·95	3
RK6	(1)	R	30·18	2
RK9	(2)	R	32·08	2
RK10	(1)	R	34·90	3
RK9	(1)	R	31·90	2
RK8	(2)	M	29·33	2
RK5	(2)	R	32·06	3
RK5	(1)	R	31·85	3
RK8	(1)	M	29·60	2
RK3	(1)	M	31·70	3
RK1	(1)	R	31·72	3
RK1	(2)	R	31·70	2
RK6	(2)	R	30·50	3
RK7	(2)	M	24·68	3

* Profiles are abbreviated as follows: CAS—Caswall Tower, CL—Cape Liddon, IN—Inland Plateau and RK—Cape Ricketts. A profile labelled CAS6 (2) indicates the second profile on the sixth slope at Caswall Tower.

† Dominant processes are abbreviated as B—basal erosion, R—rockfall and M—meltwater.

¶ The numbers indicating the form refer to the different categories mentioned in the text.

GEOMETRICAL FORM AND ROCKFALL

Rockfalls in the study area were observed in a variety of locations and during different seasons. At the start of the 1970 field season much rockfall activity had already occurred on the coastal slopes, yet air temperatures had still to reach 0°C for any lengthy period during the day. Rockfall activity continued throughout the peak melt period and the rest of the summer. It also takes place in early winter as rockfalls were observed on sunny, windless days during mid-October of 1969, a period during which maximum diurnal air temperatures did not rise above −9°C.

Cape Liddon was noticeably the most active locality for rockfalls, having both the highest section of exposed cliff face in the area (300–350 m) and the greatest amount of rockwall dissection. Caswall Tower is the second most active locality, but rockfall is confined to a much lower cliff face reaching altitudes of 175–200 m. Rockfall occurs to a lesser extent upon Cape Ricketts slopes, primarily upon talus sheets; while on the Inland Plateau slopes no recent rockfall activity is noticeable.

Along Cape Liddon in mid-June, a great deal of rockfall debris was observed at the base of the snow-covered slopes, on the beach and on the sea ice for distances up to 50 m from the shore. Many boulders with long axes in excess of 600 mm were included in the debris. The rockfall debris which accumulates on the sea ice and on the beach-fast ice is rafted away during the break-up to leave no trace of debris below the cliff section.

An important additional effect that rockfall activity has is to produce slide tongues (Rapp, 1960), usually on the steeper slopes in an area. These elongated tongues of debris produce minor variations in the overall long profile of a slope.

Twenty-nine profiles were surveyed on slopes where rockfall is dominant (Table I). Thirteen exhibited concave profiles, while twelve were classed as concave with minor convexity. The fact that twenty-five out of twenty-nine, or 86 per cent, of the rockfall-dominant slopes exhibited concavity suggests that this is the characteristic slope form produced by this process. Of the four exceptions, two slopes at Cape Liddon had straight profiles, while two convex with minor concavity profiles were measured at Caswall Tower. The continual rockfall activity over a slope appears to result in gradual extension of the slope base in both lateral and forward directions. A minor element of convexity can be introduced by the accumulation of fine-sized debris in the upper slope zone just below the rockwall base.

Slope angles for the profiles of rockfall-dominant slopes ranged from 30·2° to 37·0°, with the highest values occurring at Cape Liddon (mean slope angles 36·3°) where slope extension is limited by a narrow beach zone. Lower-angled slopes are found at Caswall Tower, Cape Ricketts and the Inland Plateau with mean angles of 33·2°, 32·5° and 32·5° respectively.

GEOMETRICAL FORM AND MELTWATER

The processes considered in this section are those that require substantial quantities of meltwater to produce slope alteration and development. Thus, we are concerned not only with surface and sub-surface flow produced by the melting of both snow and permafrost, but also with slush avalanches and debris flows. The avalanches and flows require saturation of either snow or debris by meltwater to initiate movement and produce changes in slope form. Once temperatures are consistently above 0°C, a period of rapid melting and peak water flow occurs. In 1970, the main melt period occurred between 25 June and 15 July when melting of most of the previous winter's unusually high snowfall took place. During this three-week period, the effects of meltwater were considerable on many slopes, particularly in the area between Cape Ricketts and Cape Liddon.

The natural gullies and crevices that occur in the rockwall tend to concentrate meltwater flow. As the water flows over the talus slopes it frequently causes erosion and gullying on the surface of the slope. This is particularly likely to occur when the permafrost table is at or near the surface and the meltwater cannot percolate into the slope.

As a result of the erosion, transport and deposition of material takes place. Debris from the rockwall and frost-shattered fragments from the tops of the slopes are frequently transported and deposited in cones at the base of the slopes. In some locations, deposition of large areas of fine sand and silt occurs. In both cases, the process leads to extension of the slope foot or the burial of erosion scars in the base of the slope.

Another process affecting the surface material is the slush avalanche caused by saturation of snow with meltwater. The effects of snow avalanching on slopes are clearly identified by rockfall chutes, smooth surfaces and extensive boulder aprons (Rapp, 1960). On some of the Cape Liddon slopes, many tongues of dirt and debris-laden snow indicated that this is a frequent occurrence during the initial period of meltwater flow. In one case, east of Cape Ricketts, an avalanche had descended a large gully and had carried debris over beach-fast and sea ice to a distance of nearly 100 m from the shore. As with the rockfall material, this concentrated accumulation of debris, as well as some of the debris deposited from gully erosion, was eventually rafted away on the sea ice. Thus, the occurrence of sea ice exerts a considerable influence on slope development in the coastal areas by removing debris accumulated during the period of greatest down-slope transport.

Debris mounds, resulting from saturation and transport of silt and clay occur particularly on low-angled slopes as do debris flows, involving larger sizes of material. In terms of altering the overall slope profile, however, they have little effect, particularly in comparison with the slush avalanches.

TABLE II

Mean angles of slopes grouped by locality and dominant process

| | Dominant process | | |
Location	Rockfall	Meltwater	Basal erosion
Cape Liddon	36·28°	37·60°	38·76°
Caswall Tower	33·22°	33·27°	36·87°
Inland Plateau	32·53°	20·99°	—
Cape Ricketts	32·47°	28·02°	—

Twenty profiles were measured on meltwater-dominated slopes, and for these concavity is the most frequent form recorded (Table I). Nineteen of the twenty profiles, or 95 per cent, are completely or predominantly concave, while the remaining profile has equal proportions of convexity and concavity. This exception, from Caswall Tower, has experienced slump action on its long talus foot, adding a convex element to the slope. On other slopes, slush flows and avalanches sweep debris out of the mid-slope zone, and contribute to the development of concavity by extending the basal foot.

Mean angles for these profiles reveal that the Cape Liddon slopes are steepest at 37·6°. Following in steepness are Caswall Tower, Cape Ricketts and Inland Plateau slopes with mean angles of 33·3°, 28·0° and 20·9° respectively.

ANALYSIS OF SLOPE ANGLES

The twenty-seven slope angles were grouped according to the dominant process operating (Table II) and were subjected to a series of Student t tests, parametric tests being applicable to these data (J. C. Cohen, 1965). A more desirable two-way ANOVA test could not be used since two of the slope localities have no basal erosion slopes. At the 0·01 level of significance, the angles for basal erosion slopes are greater than those of both rockfall and meltwater-dominated slopes. Rockfall slopes are also steeper than meltwater-dominated slopes at the same level of significance.

A second set of parametric *t* tests was performed on slope angles to compare the effects of different processes within each locality. At both Cape Liddon and Caswall Tower, slopes affected by marine processes are significantly steeper than rockfall slopes (at the o·05 level), but only at Caswall Tower are they also steeper than slopes affected by meltwater activity. There is no significant difference in angle between meltwater-affected and basal erosion slopes at Cape Liddon. Similarly, rockfall-dominated slopes at both localities do not differ significantly in angle from meltwater-affected slopes. These results suggest an overriding uniformity in steepness of all Cape Liddon slopes, which is attributed to the proximity of all slopes to the sea.

Both Cape Liddon and Caswall Tower have slopes close to the active sea level, and both display steep, well-dissected marine cliffs, although only part of Caswall Tower is subjected to these conditions. Cape Liddon is more exposed and susceptible to marine activity than Caswall Tower, so that the removal of material at the base by ice rafting and abrasion of debris accumulations leads to slopes with steeper angles. Evidence for overall depletion of material on some Cape Liddon slopes at the present time is seen by the recent exposure of the cliff face in the upper slope zone.

At the Inland Plateau and Cape Ricketts, where only two major process groups occur, rockfall-dominated slopes are significantly steeper than those affected by meltwater activity. In these localities, processes connected with meltwater have been more frequent in occurrence and more effective in modification of slope angles than at Cape Liddon and Caswall Tower. Avalanching and debris flow have together caused significant reductions in slope steepness.

A series of parametric Student *t* tests was performed a third time to investigate the differences if any, between slope types (sheets and cones). At the Inland Plateau and Cape Ricketts areas, sheet angles are significantly steeper than cone angles (at the o·01 level). This reflects the results obtained above in that all talus cones have been affected by some type of water activity, while a great majority of the sheets sampled are rockfall-dominant slopes. At Caswall Tower, cone angles are significantly steeper (at the o·01 level) than sheet angles, but here it is only the cones that are dominated by basal erosion, which would account for the difference. Cape Liddon slopes similarly reflect variations of process in that sheet- and cone-angles do not differ significantly. Previous tests found that only basal erosion slopes were steeper here than others but the fact that the slopes affected by basal erosion consist of both sheets and cones accounts for the lack of differentiation in steepness.

The results of the parametric tests indicate that differences in steepness between slope types reflect the differences found between the effects of dominant processes.

DISCUSSION

Comparisons have been made between these results and those from previous studies. The mean angles for rockfall slopes in Table II agree with W. H. Ward's (1945) 30°–35° range for dry, gravitational slopes in the south of Britain and those for slopes in Labrador (J. T. Andrews, 1961). K. J. Tinkler's (1966) limiting angle of 36° for such slopes corresponds with the highest mean recorded for rockfall slopes in south-west Devon Island. These values are slightly lower than C. H. Behre's (1933) 36·5° characteristic angle in alpine areas, but this is related to the centralizing tendencies of producing a mean for a locality. Such comparisons show that only slight variations in angles arise as a consequence of environmental differences, and that the characteristic angle for rockfall slopes is about 36°.

Comparisons were also made between Spitsbergen slopes (Rapp, 1960) and those of south-west Devon Island affected by similar processes. In all cases, testing by parametric ANOVA has indicated no significant differences between each set of angles. The Spitsbergen values are slightly higher than those of the study area in all cases and might be regarded as the limiting angles for each set. For rockfall slopes, mean angles range from 33·5°–35·5°; while for meltwater-affected and basal erosion slopes the ranges are 30·3°–31·0° and 38·0°–40·0° respectively. For both Spitsbergen and south-west Devon Island, parametric ANOVA revealed a significant difference (at the 0·05 level) between angles of slopes affected by different processes.

Results of the profile analysis of slopes dominated by the three different processes have been given earlier. To summarize, it is found that twenty-five out of twenty-nine, or 86 per cent, of the rockfall slope profiles are completely or predominantly concave in form, while convexity dominates two out of five, or 40 per cent, of the basal erosion slope profiles. Slopes where meltwater activity is dominant are the most pronounced in form, since 95 per cent (nineteen out of twenty) are completely or predominantly concave with the one remaining profile convexo-concave.

In Spitsbergen, Rapp (1960) observed that talus cones with slide tongues tend to be straight in form, while those with avalanches and mudflows are reported to be distinctly and slightly concave respectively. There appears, then, to be agreement as to profile form associated with various processes in active High Arctic periglacial environments.

CONCLUSIONS

Studies have been made of three different processes affecting debris slopes on south-west Devon Island. The results indicate that in most areas significant differences in geometrical form occur between slopes subjected to different dominant processes. It is possible to explain the differences based on observations of the processes operating in the field. Slopes subjected to similar processes in both High Arctic periglacial and mid-latitude environments do not differ significantly in the slope angles which they display. The debris slope angles measured on south-west Devon Island are also similar to those obtained by Rapp (1960) on Spitsbergen.

DISCUSSION

B. A. KENNEDY: A less complex measure of slope form might be the height/length integral (R. J. Chorley and Kennedy, 1971) derived from A. N. Strahler's (1952) hypsometric integral. This not only gives a direct measure of convexity or concavity, but can be compared directly with M. J. Kirkby's (1971) characteristic curves for slopes dominated by different processes. The field observation of three different classes of scree slope in terms of dominant process could then be checked against Kirkby's theoretical forms.

J. B. THORNES: Could the authors say more precisely how they obtained the mean slope angle? The questioner doubted the value of using a mean slope angle to represent convex and concave slope sections and also wondered whether, by standardizing slope length, the authors were ignoring one important variable as far as process is concerned.

P. J. HOWARTH and J. G. BONES: Use of the height/length integral and a consideration of Kirkby's (1971) ideas might prove useful in future work. This paper was submitted, however, before the book by Chorley and Kennedy (1971) was published. The designation of convex and concave sections was used as an objective measure to describe and classify the types of slope that occurred. It is the proportions of each type of section that are important in this classification rather than slope angle. The survey method was selected to provide rapid measurement on a reasonable number of slopes, but as pointed out, it only permitted the major variations in slope form to be shown as straight-line segments. The limitations presented by this method when plotting profiles and determining slope angles are realised. Standardization of slope length, though perhaps losing information, was necessary in this analysis to provide a dimensionless number. (For References quoted in discussion, see p. 153. Ed.)

ACKNOWLEDGEMENTS

The authors are pleased to acknowledge financial support received from the National Research Council of Canada, the McMaster University Research Board and the National Advisory Committee for Geographical Research.

REFERENCES

ANDREWS, J. T. (1961) 'The development of scree slopes in the English Lake District and Central Quebec-Labrador', *Cah. Géogr. Québ.* 5, 219–30

BEHRE, C. H. (1933) 'Talus behaviour above the timberline in the Western Rockies', *J. Geol.* 41, 622–35

BIRD, J. B. (1959) 'Recent contributions to the physiography of Northern Canada', *Z. Geomorph.* 3, 150–74

BIRD, J. B. (1967) *The Physiography of Arctic Canada*

BONES, J. G. (1971) 'Process-morphology interaction on Arctic debris slopes, south-west Devon Island, Canada', Unpubl. M.Sc. thesis, McMaster University

COHEN, J. C. (1965) 'Some statistical issues in psychological research' in *Handbook of clinical psychology* (ed. B. B. WOLMAN), 95–121

FORTIER, Y. O. *et al.* (1963) 'Geology of the north-central part of the Arctic Archipelago, N.W.T. (Operation Franklin)', *Mem. geol. Surv. Brch Can.* 320

JAHN, A. (1968) 'Morphological slope evolution by linear and surface degradation', *Geogr. Polonica* 14, 9–21

McCANN, S. B. and E. H. OWENS (1969) 'The size and shape of sediments in three Arctic beaches, south-west Devon Island, N.W.T., Canada', *Arct. alp. Res.* 1, 267–78

OWENS, E. H. and S. B. McCANN (1970) 'The role of ice in the Arctic beach environment with special reference to Cape Ricketts, south-west Devon Island, N.W.T., Canada', *Am. J. Sci.* 268, 397–414

RAPP, A. (1960) 'Talus slopes and mountain walls at Tempelfjorden, Spitsbergen', *Skr. norsk Polarinst.* 119

STOCK, R. F. (1968) 'Morphology and development of talus slopes at Ekalugad Fiord, Baffin Island, N.W.T.', unpubl. B.A. thesis, Univ. of Western Ontario

THORSTEINSSON, R. (1958) 'Cornwallis and Little Cornwallis Islands, District of Franklin, N.W.T.', *Mem. geol. Surv. Brch Can.* 294

TINKLER, K. J. (1966) 'Slope profiles and scree in the Eglwyseg valley, North Wales', *Geogrl J.* 132, 379–85

WARD, W. H. (1945) 'Stability of natural slopes', *Geogrl J.* 105, 170–97

YOUNG, A. (1964) 'Slope profile analysis', *Z. Geomorph.* Suppl. 5, 17–27

RÉSUMÉ. *Rapports entre les processus et la géométrie des formes le long des pentes d'accumulation de débris dans le Haut-Arctique, au Sud-ouest de l'île de Devon, Canada.* Le profile des pentes d'accumulation de débris a été mésuré en quatre endroits différents au sud-ouest de l'île de Devon. L'article décrit la nature et les effets de l'érosion de base, des éboulis et des processus conditionnés par les eaux de fonte. L'analyse des angles de ces pentes indique qu'il y a des différences significatives entre les angles de pentes formées par des processus différents. Les pentes les plus raides, de forme généralement convexe, sont le produit d'une érosion de base résultant de poussées de glace et de l'action de vagues intempestives. À distance de la côte on a découvert que les pentes où dominent les éboulis sont sensiblement plus prononcées que celles qui furent produites par l'action des eaux, bien que ces deux soient caractérisés par des formes généralement concaves. On peut également observer les effets de ces deux processus dans la différence marquée entre les nappes de talus (dues aux éboulis) et les cônes de talus (dûs à l'action des eaux de fonte). Dans les zones côtières, cependant, les pentes dues aux éboulis de même que celles qui proviennent d'avalanches sont prononcées du fait que l'accumulation à la base n'a pas eu lieu, les débris ayant été dispersés par les radeaux de glace. Il semble donc qu'il n'existe que peu de différence dans la géométrie des formes entre les pentes du Haut-Arctique et celles des latitudes moyennes, si l'on compare les angles et les profiles qu'on peut rattacher aux mêmes types de processus de formation.

FIG. 1. Sites des pentes étudiées dans la région de la baie de Radstock. Des dépôts de plage occupant la majeure partie de reste du terrain

FIG. 2. Une plateforme rocheuse au cap Liddon exposée par l'érosion de base

FIG. 3. Glace et permafrost exposés à la base d'une pente d'accumulation par l'érosion de base

FIG. 4. Le mouvement de matériaux vers le base de la pente a recouvert la plateforme rocheuse visible sur la Figure 2

ZUSAMMENFASSUNG. *Zusammenhänge zwischen Entstehungsprozess und geometrischer Form von hocharktischen Geröllhalden, Süd-West Devon Insel, Kanada.* Messungen von Geröllhalden-Profilen wurden an vier verschiedenen Stellen auf der Süd-West Devon Insel, N.W.G., Kanada, angestellt. Für diese vier Stellen wurden Art und Auswirkungen von Basis-Erosion, Steinfall und Schmelzwasser beschrieben. Eine analyse des Haldenwinkel ergibt, dass bedeutende Unterschiede im Winkel zwischen von verschiedenen Prozessen gebildeten Halden bestehen. Die steilsten Halden, die zu konvexen Formen neigen, stammen von Basis-Erosion, die sowohl auf Eisstoss und Sturmwellen zurückzuführen ist. An Küstenfernen Standorten wurde festgestellt, dass vorwiegend von Steinfall gebildete Halden bedeutend steiler sind als die von Wasser verursachten, obwohl beide Haldenarten zu konkaver Form neigen. Die Auswirkungen dieser beiden Prozesse sind auch an den wesentlichen Unterschieden zwischen Schutthalden (vorwiegend Steinfall bedingt) und Schuttkegeln (vorwiegend Schmelzwasser Zedingt) zu sehen. In Küstengebieten jedoch sind sowohl Steinschlag—als auch Lawinenhänge steil, da das Geröll von Treibeis weggeschwemmt wird um Basisansammlungen zu vermeiden. Zwischen hocharktischen Halden und jenen mittlerer Breiten scheinen geringe Unterschiede in der Form zu bestehen wenn Winkel und Profile von Hängen verglichen werden, die von ähnlichen Prozessen in den beiden Gebieten gebildet wurden.

ABB. 1. Lage der untersuchten Schutthalden im Gebiet der Radstock-Bucht. Die meisten übrigen Gebiete bestehen aus Küstenablagerungen

ABB. 2. Eine von Basis-Erosion freigelegte Felsplatte auf Kap Liddon

ABB. 3. Eis und Dauerfrostboden am Fusse einer Schutthalde, freigelegt durch Basis-Erosion

ABB. 4. Abgerutschtes Material hat die in Abb. 2 gezeigte Felsplatte überdeckt

---

REFERENCES QUOTED IN DISCUSSION

CHORLEY, R. J. and B. A. KENNEDY (1971) *Physical geography: a systems approach*

KIRKBY, M. J. (1971) 'Hillslope process-response models based on the continuity equation', *Inst. Br. Geogr. Spec. Publ.* 3, 15–30

STRAHLER, A. N. (1952) 'Hypsometric (area-altitude) analysis of erosional topography', *Bull. geol. Soc. Am.* 63, 1117–42

Processes of soil movement in turf-banked solifluction lobes, Okstindan, northern Norway

C.HARRIS

(Tutor in Geography, University College of Swansea)

Revised MS received 10 November 1971

ABSTRACT. The total annual rates of soil movement on an east-facing slope of between 4° and 20° were measured, together with soil-moisture conditions, winter frost heave, soil temperatures, the vertical profile of soil movement with depth, and the mechanical properties of the soils themselves. Utilizing these data, the nature of soil movement is discussed.

DOWNSLOPE soil movement in Alpine and Polar regions has attracted considerable attention in the past few years, and there have been many measurements of rates of movement. Studies such as those by P. J. Williams (1959), A. L. Washburn (1967), and J. B. Benedict (1970) have attempted to take into account environmental conditions and assess their effect on the measured rates of movement, and it was with such an approach in mind that this project was commenced. Soil properties and climatic conditions within the study area were measured in addition to the rates and depth characteristics of soil movement. From these data an attempt was made to analyse the mechanics of soil movement. Results presented in this paper are of an interim nature, since the project is necessarily long-term, and further data on annual rates of soil movement are to be collected.

PROCESSES OF SOIL MOVEMENT

It is generally accepted that down-slope soil movement in periglacial environments results from two major processes, frost creep and solifluction. Frost creep was defined by Washburn (1967) as the 'ratchet-like down-slope movement of particles as the result of frost heaving of the ground and subsequent settling on thawing, the heaving being predominantly normal to the slope and the settling more nearly vertical'. Soil creep can also result from expansion and contraction of the soil because of wetting and drying (A. Young, 1960) and this may contribute considerably to creep in periglacial areas where the soil is non-frost-susceptible. J. G. Andersson (1906) defined solifluction as the 'slow flowing from higher to lower ground of masses of waste saturated with water'. In the field it was not possible to differentiate between creep and solifluction, the total annual down-slope soil movement being recorded.

THE STUDY AREA

The Okstindan mountain range is situated some 30 km to the south-east of Rana Fjord in Nordland, north Norway (Fig. 1) and supports an ice cap which extends across latitude

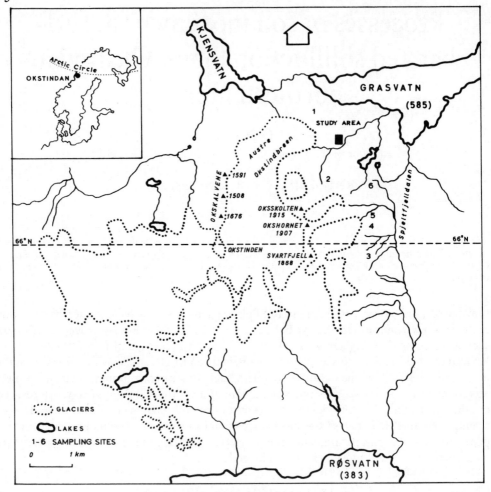

FIGURE 1. Location map

66°N, covering an area of approximately 60 km². The highest peak of the range, Oksskolten, is ice-free, rising to 1915 m, whereas the ice cap itself lies at a general height of around 1450 m. The eastern part of the area is made up of mica schist. The Okstindan range lies only about 80 km from the sea and therefore experiences a relatively maritime climate. Hattfjelldal, the nearest inland Norwegian meteorological station, is situated about 50 km to the south at an altitude of 221 m. Mean January and July temperatures for this station are −9·9°C and 12·4°C respectively (1954–69) and the mean precipitation is 1120·6 mm, with an average of 181 days of snow cover.

Frost creep and solifluction appear to be important agents in slope evolution within the Okstindan area, and both turf-banked and stone-banked terraces and lobes are common. It was decided to make a detailed study of a specific area where these processes of down-slope soil movement appeared to be active. The area chosen for study lies on the north-eastern flank of the range at an altitude of around 710 m (Fig. 1). It consists of an east-facing slope

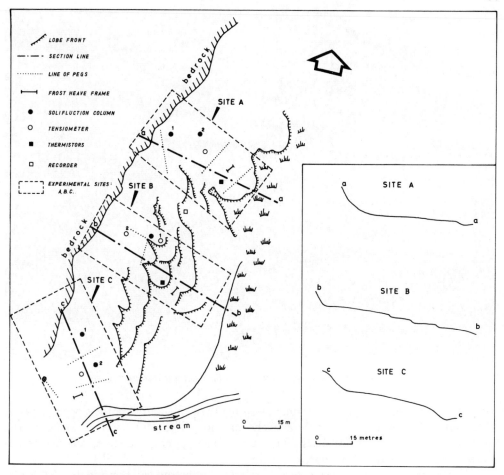

FIGURE 2. The study area, showing location of experimental sites and their instrumentation together with slope profiles of experimental sites

of between 5° and 17°, bounded at the top by a sharply rising mica-schist roche moutonnée, and at the bottom by a flat marshy area. The slope is formed on a sandy till deposited against the roche moutonnée.

Except in the southern one-third of the area, well-developed turf-banked solifluction lobes extend down the slope, with frontal heights ranging from 0·5 m to 1·5 m, and axial lengths from 10 m to 50 m (Figs. 2 and 3). Preliminary reconnaissance in 1968 suggested that soil movement was active on this slope and, with bedrock at its head providing for fixed survey stations, the area appeared suitable for a study of rates and processes of soil movement.

A distinct vegetational pattern is apparent throughout most of the study area, the upper section of the slope being moss-covered, giving way down-slope to Least Willow (*Salix herbacea*) which in turn gives way to a heath community dominated by Bearberry (*Arctostaphulos alpina*). This vegetational pattern is determined by winter snow accumulation. Snow drifts in the lee of the roche moutonnée and consequently is thickest at the head of

FIGURE 3. Part of the study area. Height of the risers approximately 50 cm

the slope, thinning down-slope. During the spring thaw the bottom of the slope is exposed first and snow cover retreats progressively up-slope. The drift survives well into the summer as a late-lying snow patch. The vegetational pattern, therefore, reflects the decrease in snow-free days up-slope.

Within the study area three plots were selected for instrumentation and detailed study. These are referred to as sites A, B and C. Site A consists of a large lobe with frontal height 1·5 m and axial length 50 m; B consists of a series of smaller 'garland lobes'; and C, in the southern extremity of the area, consists of a smooth convex slope which does not exhibit lobate forms.

SOIL PROPERTIES

Grain size

Samples were taken from each of the experimental sites for mechanical analysis. Pits dug to accommodate frost-heave frames (A. Jahn, 1961) and thermistors were utilized as sampling locations, and in each site samples from 10, 50 and 100 cm below the surface were taken from two locations.

Results are shown as cumulative frequency curves (Fig. 4). In very general terms the soils in sites A and B may be described as fine silty sand and in site C as gravelly sand. In the case of A and B, the grain-size properties indicate soils which are frost-susceptible according to G. Beskow (1935), while at C the higher percentage of sand and gravel

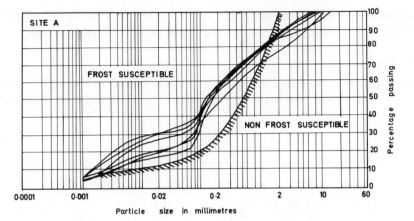

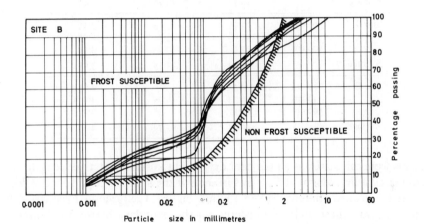

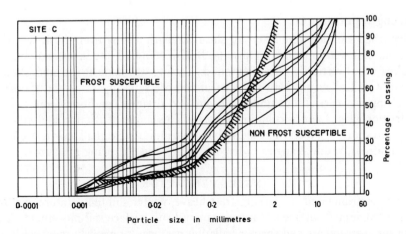

FIGURE 4. Grain-size composition; samples taken from depths of 10, 50 and 100 cm below the surface at two locations in each site. Note that the curves include the fraction coarser than 2 mm

suggests a somewhat less frost-susceptible soil. Plotting the fraction finer than 2 mm against Beskow's limits indicates that this site is marginally frost-susceptible.

Atterberg limits

The index properties were found to be fairly constant in each of the experimental sites. They are summarized in Table I. It should be noted that the saturated moisture contents

TABLE I
Atterberg limits

Location	Depth	Liquid limit %	Plastic limit %	Plasticity index %
Site A	10 cm	23·0	21·9	1·1
	50 cm	21·2	19·7	1·5
	100 cm	24·0	19·25	4·75
Average		22·73	20·28	2·45
Site B	10 cm	26·7	23·15	3·55
	50 cm	25·08	23·0	2·08
	100 cm	21·4	17·9	3·5
Average		24·39	21·35	3·04
Site C	10 cm	20·75	18·1	2·65
	50 cm	20·75	16·75	4·0
	100 cm	20·5	18·5	2·0
Average		20·67	17·78	2·89

of these soils were calculated as 24 per cent dry weight in site A, 25·3 per cent dry weight at B, and 29 per cent dry weight at C. Thus in each of the sites the liquid limit is lower than the saturated moisture content. When plotted on the Casagrande Plasticity Chart (A. Casagrande, 1932a), these soils lie clustered on the boundary between cohesionless soils and inorganic silts of low compressibility.

Shear-strength parameters

Two undisturbed samples were taken for shear-strength testing from the area to the south of B. The samples were obtained by driving British standard site investigation 4-inch (10 cm) sampling tubes vertically into the ground. Testing was done at the laboratories of the Cementation Co. Ltd, using both the triaxial compression apparatus and the shear box. Triaxial testing was carried out under consolidated drained conditions, and indicated an angle of friction of 35° and a cohesion of zero. This angle of friction and cohesion was verified by two shear-box texts. The shear-box tests indicated a small loss in strength under residual conditions, residual angles of friction of 29° and 33° being measured.

INSTRUMENTATION

The area was instrumented to measure frost heave, winter soil temperature, soil moisture conditions, total annual surface soil movement and sub-surface soil movement. A detailed account of the instruments and their installation has been previously reported (C. Harris, 1971). The locations of the instruments described below are shown in Figure 2.

Frost heave

The maximum winter frost heave and summer resettling was recorded in each of the experimental sites using frost-heave frames similar in design to those used by Jahn (1961). In all three sites the initial frost heave was measured relative to the ground surface in October 1969. In each case the subsequent resettling during the summer of 1970 exceeded the recorded frost-heave value for the winter of 1969–70 (Table II). This discrepancy was related to drying of the soil during the summer of 1970. In October 1969, soil moisture contents of around 20 per cent dry weight were recorded, compared with less than 10 per cent dry weight in August 1970. Thus the total vertical displacement of the ground from minimum elevation was caused partly by swelling as a result of increased moisture content in the autumn, and partly by the freezing of the soil during winter.

TABLE II

Frost heave and potential frost creep, 1969–70 and 1970–71

Location	Slope angle	Winter frost heave 1969–70 cm	Summer resettling 1970 cm	Potential frost creep 1969–70 cm	Winter frost heave 1969–70 cm	Summer resettling 1971 cm	Potential frost creep 1970–71 cm
Site A	Upper slope 17°	2·8	3·5	1·06	9·2	8·8	2·57
	Lower slope 5°	2·8	3·5	0·35	9·2	8·8	0·76
B	15°	0·7	1·8	0·48	2·3	2·1	0·54
C	13°	0·65	0·8	0·18	?	?	?

It has been shown by S. Taber (1930), Beskow (1935), A. Higashi (1958) and others that the rate of frost heaving depends largely on the grain size of the soil, its moisture content and its rate of freezing. The lower value of frost heaving recorded at C was to be expected, since the soil was only marginally frost-susceptible. The disparity between the values recorded at A and B, however, must be explained by their rates of freezing since their grain size characteristics and moisture contents were similar.

The lower slopes at B were kept largely clear of snow by wind scour during the winter whereas accumulation occurred in the more sheltered area of A. Here the snow thickness measured against a stake adjacent to the frost-heave frame showed snow depths of 72 cm in January, 85 cm in April and 78 cm in May 1970, and 130 cm in April 1971, compared with a discontinuous snow cover of maximum thickness 5 cm in the area of the frost-heave frame at B, throughout both the winter of 1969–70 and that of 1970–71. This suggests that soil freezing was probably much slower at A owing to the insulating effect of the snow, a suggestion which was confirmed by the recorded soil temperatures in the autumn of 1970. The much greater frost heave recorded at A reflects the difference in the rates of soil freezing between the two sites. A slower penetration of the freezing plane in the winter of 1970–71 may also explain the much greater frost-heave values recorded at both A and B in that year (Table II). Results of frost heave at C for the winter of 1970–71 are not available since the frost-heave frame in this site was damaged by the weight of overlying snow and yielded no results.

Factors which may have been responsible for a slower rate of freezing in the winter of 1970–71 include higher air temperatures during that winter, snow cover present over a longer period during the autumn of 1970 than during the autumn of 1969, and snow depth greater during the winter period of 1970–71 than during that of 1969–70. There is no continuous record of air temperatures in the study area through the winter of 1969–70 and 1970–71. However, a comparison between the air temperatures recorded at Hattfjelldal during the two winters under consideration indicates the relative severity of these two winters, even though the Hattfjelldal temperatures were probably higher than those in the study area owing to the altitudinal difference. Table III shows the mean monthly air

TABLE III

Air temperatures at Hattfjelldal, September–May

	Mean monthly air temperature 1969–70 °C	Mean monthly air temperature 1970–71 °C
September	5·5°	6·2°
October	2·8°	1·5°
November	−7·4°	−5·8°
December	−8·4°	−3·7°
January	−11·6°	−5·2°
February	−13·6°	−6·9°
March	−3·7°	−8·2°
April	−1·5°	−0·7°
May	6·7°	5·3°

temperatures recorded at Hattfjelldal through the winters of 1969–70 and 1970–71. The mean monthly air temperatures in the second winter were considerably higher than those of the first and, consequently, the rate of soil freezing must have been slower in this second winter than was the case in the first. This led to the development of thicker ice lenses within the soil during the winter of 1970–71, and hence to much greater frost heave of the soil surface.

As with air temperatures, no record is available of the number of days with snow cover in the study area, but the record from Hattfjelldal suggests a greater incidence of snow cover in the autumn of 1970 (September to December, inclusive) than in the autumn of 1969, with 62 days of snow cover in 1970 compared with 41 days of snow cover in 1969. Assuming that this greater incidence of snow cover in 1970 was general throughout Okstindan, it appears that the soil was insulated from sub-zero air temperatures by snow cover over a longer period during the autumn of 1970 than that of 1969, and this may well have been a further factor in slowing the penetration of the freezing plane through the soil, thereby producing the greater frost-heave values of the winter of 1970–71.

In addition to the duration of snow cover, the thickness of snow also has an effect on the rate of heat loss from the soil, the thicker the snow mantle, the greater being its insulating effect. Snow-depth readings have already been given, and although only one reading was taken for the winter of 1970–71, in April, this showed significantly more snow at A than at the same time in the previous year, though wind action had maintained relatively snow-free conditions at B. Bearing in mind the greater incidence of snow cover recorded at Hattfjelldal, it seems likely that snow cover was generally thicker in the winter of 1970–

71 than in the previous year. Again the effect of a greater thickness of snow would have been to retard the freezing of the soil during the winter of 1970–71.

It appears, therefore, that all three factors, air temperature, period of snow cover and snow depth, may have contributed to producing the higher values of frost heave recorded during the second winter of this study.

Higher soil moisture content during the winter freeze of 1970–71 may also have contributed to the greater frost heave measured in that year. Soil moisture conditions measured during the summer of 1970 indicated that, following snow clearance in early summer, soil moisture content depended largely upon precipitation. Summer and autumn precipitation at Hattfjelldal shows a total of 473 mm in 1969 (July to November inclusive) compared with 448 mm in 1970. These figures cannot be taken as representing actual values of precipitation in the study area but serve as an approximate means of comparing precipitation in these two summers. The difference in precipitation was not large and, assuming that this pattern of precipitation was maintained 50 km to the north in the study area, it may be concluded that soil moisture conditions were roughly similar at the end of the summer periods of 1969 and 1970, and that increased soil moisture content during freezing does not appear to have been a factor in producing the greater frost-heave values recorded in 1970–71.

Whatever the cause of the variations recorded, however, this experiment serves to illustrate the need for long-term measurement of frost heave within any particular area before generalizations may be made concerning average values. The values recorded in this limited study showed a variation of up to 300 per cent from one year to the next in one particular location, and there is no evidence to suggest that this was exceptional. The experiment also demonstrates the large variation in frost heave that may occur during one winter within a relatively small area. The high degree of snow drifting within Okstindan appears to be a major contributing factor to this spatial variation.

From the recorded values of frost heave and the slope angles, the potential soil creep was calculated for each experimental site, assuming a uniform frost heave at each site (Table II). The maximum potential soil creep occurs when the soil resettles vertically, and is given by $C = h \tan \alpha$ where $C =$ creep, $h =$ the amount of resettling following heave, and $\alpha =$ angle of slope.

Winter soil temperatures

It was planned to measure the winter soil temperatures using thermistors and an automatic temperature recorder[1]. The thermistors were placed at depths of 100, 50, 25 cm and at the surface at A, and 75, 25 and 5 cm at B; one was also placed in a small white aerated box 1 m above the ground to measure air temperature. However, the recorder proved rather unreliable. It worked for only 4 weeks after the field party left in October 1969; it was restarted in January 1970 but worked for only a few hours, and behaved similarly in April. In May, with the assistance of a new battery, it was restarted and worked satisfactorily until mid-June, by which time the soil had thawed completely.

It was mentioned above that there was considerable contrast in snow accumulation between A and B, and this was reflected in the soil temperatures recorded in January and April 1970 (Table IV). The insulating effect of the snow cover at A produced winter soil temperatures several degrees higher than at B. Soil temperatures through the spring thaw of 1970 are shown in Figure 5. It is interesting to note that, at the 100 cm depth in site A, the soil thawed 20 days earlier than at the 50 cm depth, indicating that melting of the soil

TABLE IV

Winter soil temperatures, 1970

Depth cm	Site A 25 January °C	Site A 4 April °C	Site B 25 January °C	Site B 4 April °C
0	−3·5°	−5·0°	–	–
5	−2·25°	−2·8°	−12·6°	−15·3°
25	−2·1°	−1·3°	−9·5°	−11·2°
50	−1·5°	−1·2°	–	–
75	–	–	–	−3·5°
100	−0·3°	−0·6°	–	–

January air temperature −28°C   April air temperature −12·4°C

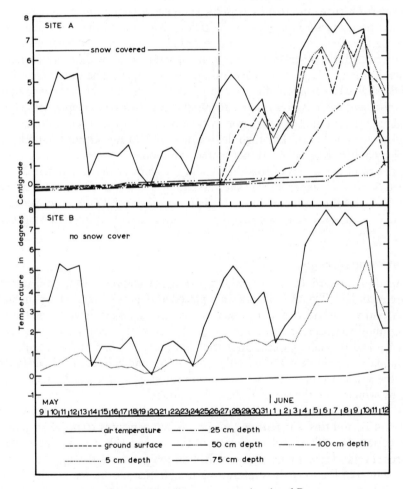

FIGURE 5. Spring soil temperatures, sites A and B, 1970

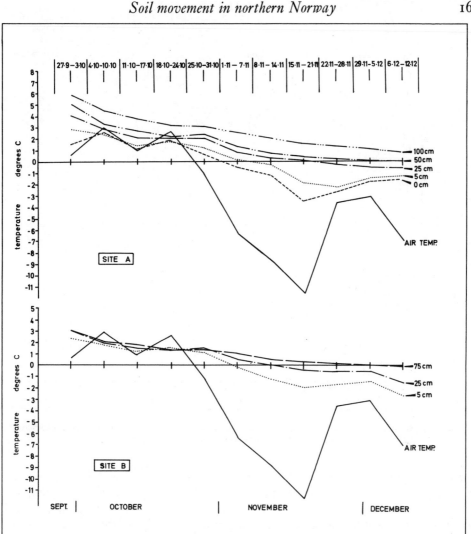

FIGURE 6. Autumn soil temperatures, sites A and B, 1970

was taking place from below as well as from the surface. This suggests that geothermal heat as well as solar heat was responsible for thawing of frozen ground in this area.

During the summer of 1970 the temperature recorder was returned to the manufacturers and modified to give only six readings per day, in place of the previous twelve readings per day. It was hoped that this would place less of a strain on the battery and therefore improve the performance of the recorder. The recorder was returned to the field in late summer and was installed on 27 September 1970. Records of soil temperatures were obtained from this date until 12 December when the battery apparently failed and the recorder ceased to operate. The data obtained over this period provided useful information concerning soil freezing during the autumn of 1970.

The general cooling of the soil is best shown by the weekly mean temperatures plotted in Figure 6. From the dates at which each monitored depth within the soil reached 0°C, the

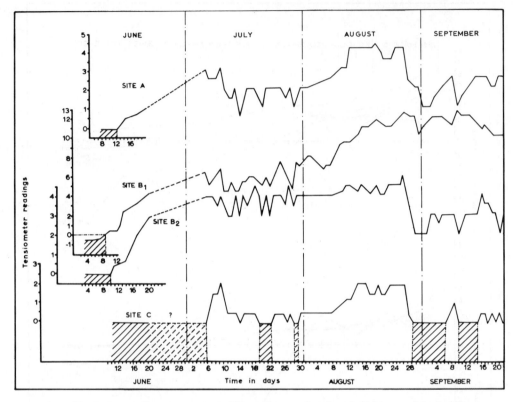

FIGURE 7. Tensiometer readings, summer 1970. The readings are given in scale divisions, and the shaded areas under the graphs show periods of soil saturation.

average rates of penetration of the freezing plane were calculated, at A from the surface to 50 cm depth, and at B from 5 cm to 75 cm depth. An average rate of soil freezing of 1·23 cm/day was found at A, compared with 2·33 cm/day at B. Thus soil freezing took place almost twice as fast at B as at A, reflecting the greater insulation of the soil surface by snow at A. The significance of this large difference in rates of soil freezing to the amount of winter frost heaving has already been discussed.

Soil moisture conditions

Porous-pot soil moisture tensiometers were used to monitor changes in soil moisture conditions through the summer of 1970. Two tensiometers were installed at each of the experimental sites (Fig. 2), but one tensiometer at A and one at C proved to be faulty and yielded no results. The tensions registered on the remaining instruments were not directly comparable owing to differences in the efficiency of their vacuum gauges, and each tensiometer was therefore subsequently calibrated in the laboratory against a water manometer tensiometer. Field results are shown in Figure 7.

Site A Zero pore-water tension (saturated conditions) was indicated from the time of installation until 13 June when a tension was first registered. Drying of the soil continued from this date. Reference to the spring soil temperatures of 1970 (Fig. 5) shows that the frozen subsoil at A, in the area of the thermistors, melted completely on 6 June, some 10 days after the snow cleared from that area. Snow clearance in the area of the tensiometer took place on 4 June, and assuming slightly faster thawing of the soil in this area as a result of rising air temperatures, it appears reasonable to suggest that thawing of the soil took place between 4 and 13 June; during this time, drainage was impeded by ground ice, resulting in meltwater saturating the unfrozen surface soil layers.

Site B Both tensiometers indicated initially saturated conditions within the soil, tensiometer 1 until 8 June and tensiometer 2 until 10 June. Tensiometer 1 in fact registered a subzero tension, suggesting a pore-water pressure at the depth of the porous pot.

The soil temperature record for B showed complete soil thawing in the area of the thermistors by 12 June. Thus there is no conclusive evidence that drainage of the higher portions of the slope in the region of the tensiometers coincided with the melting of a frozen subsoil. However, the Bearberry heath vegetation in the area of the thermistors was much thicker than the moss-Least Willow vegetation around tensiometer 1, providing greater insulation from the rising spring temperatures. Thawing could therefore have taken place more rapidly in the latter area, resulting in melting of the subsoil 4 days earlier than farther down-slope.

Thawing in the area of tensiometer 2 began with the clearance of the snow cover on 31 May, since this higher area was affected by the snowdrift against the roche moutonnée. The presence of this snow cover, as at A, would have prevented deep penetration of the freezing plane, and it therefore appears reasonable to suggest a similar period for soil thawing as at A, that is, 9 to 10 days, producing clearance of ground ice by 10 June.

It is suggested, therefore, that desaturation of both tensiometers coincided with the final thawing of the frozen subsoil, which had previously impeded drainage through the soil. Again meltwater from the snowpatch maintained saturated soil conditions until complete melting of the subsoil took place.

Site C Snow clearance at C took place later than elsewhere, and the snowpatch survived until late July. Zero tension was recorded until 20 June, when readings were temporarily suspended, and when recommenced on 6 July, zero tension was again recorded. Throughout the summer, saturated soil conditions were re-established following periods of precipitation, and it was apparent that soil saturation in this area did not depend on impedance of drainage by ground ice. The high incidence of soil saturation is explained by a much greater supply of moisture to the area than at A and B. In addition to meltwater from the snow patch, run-off both from snow melt and precipitation over much of the roche moutonnée flowed down over the surface at C, producing saturation during the spring thaw, and following precipitation during the summer.

Total annual surficial soil movement

It was planned to measure annual surface movement using dowel pegs, 1·5 cm in diameter and 30 cm long, inserted at 2 m intervals across individual lobes (Fig. 2). They were inserted to alternate depths of 10 cm and 20 cm. Theodolite stations were established on the roche moutonnée, and the pegs were surveyed by triangulation from these stations.

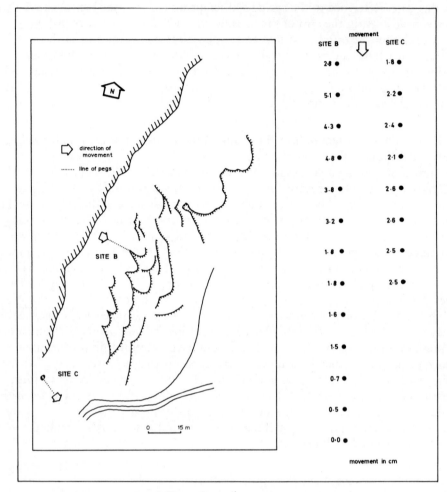

FIGURE 8. Down-slope soil movement, 1970–71

However, this survey method proved insufficiently accurate to measure the peg displacements. Peg movements calculated from the theodolite readings showed variations far in excess of that possible within this area.

A line of pegs, parallel to the line of greatest slope, was set up during the summer of 1970 in each of sites B and C. These were designed to measure near-surface soil movement and were inserted to only 5 cm depth. Their displacement was measured with a tape with reference to a stake sunk to a depth of 75 cm at B and to a large boulder of diameter approximately 4 m at C. The boulder was considered to be stable since it was embedded within the soil to a depth of over 1 m, and soil appeared to be moving around it.

From these lines of pegs, data on movement for the year 1970–71 were obtained for sites B and C and are shown in Figure 8. Data on soil movement for the year 1969–70 were provided by the results of the test pillars installed in each of the experimental sites in 1969 to measure sub-surface soil movement (Table V).

Sub-surface soil movement

One of the simplest techniques for studying sub-surface soil movement is to bury vertical columns comprising short lengths of material within the soil. Re-excavation after a period of a year or more will then reveal the amount of displacement of the individual column members by the moving soil at various depths (Williams, 1957; S. Rudberg, 1958; Benedict, 1970). In the present study polythene tubing of external diameter 2·0 cm and internal

TABLE V

Summary of soil movement measured from the buried columns, 1969–70

Location	Surface movement 1969–70 cm	Depth of movement 1969–70 cm	Angle of slope
Site A Column 1	3·5	40	17°
Column 2	1·75	20	5°
B	1·0	30	15°
C Column 1	6·0	35	11°
Column 2	4·0	30	6°

diameter 1·6 cm was cut into 5 cm lengths. These were threaded over a straight rod of slightly smaller diameter than the internal diameter of the tube sections. A vertical auger hole was then made and the column of tube sections mounted on the rod was pushed into the hole. The rod was then removed, leaving the column of polythene segments intact within the soil. After one year (a longer period would have been preferred), the columns were carefully re-excavated and the displacement of the sections affected by soil movement was measured against a plumb line (Fig. 9). In the upper section of the slope at C, a polythene tube of external diameter 1 cm was inserted into the soil, using the same method of insertion as for the columns, described above. This was re-excavated in the same way as the columns, to reveal the characteristics of soil movement with depth in that location. The surface soil movement and the depth of movement indicated by these columns is shown in Table V.

The rates of soil movement

At C the high moisture content for much of the spring and early summer of 1970 was reflected in a surface movement of 6 cm on a slope of 11°, and 4 cm on a slope of 6°, the two largest recorded surface soil movements in the study area for the year 1969–70. At A, on a slope of 17°, a down-slope movement of only 3·5 cm was recorded. Clearly the angle of the slope is of secondary importance compared with the soil moisture conditions in controlling down-slope rates of soil movement within this area. The effect of a thick cover of vegetation acting as a retarding 'skin' can be seen at B, the surface movement being less than that immediately below the surface.

Peg movement for the year 1970–71 at B was greatest near the head of the slope (Fig. 8). Here the slope was affected by the late-lying snow patch and was moss-covered. Farther down-slope, where wind maintained relatively snow-free conditions during winter, Least Willow (*Salix herbacea*) and Bearberry (*Arctostaphulos alpina*) became established and this thicker vegetation resulted in a considerable reduction in peg movement.

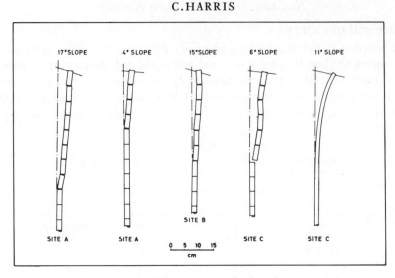

FIGURE 9. Soil movement at depth, 1969–70

At C the snow patch covered the whole of the slope upon which the peg line was established. Here the vegetation consisted of moss and Least Willow. The slope angle was constant at around 10° and the resulting peg movement remarkably uniform. This line of pegs lies to the south of the zone of maximum soil movement at C, suggested by the movement values recorded by the buried columns, and was not affected by the surface run-off from the roche moutonnée. It appears, therefore, that the period of soil saturation in this southern section of site C was much shorter than in the central section, where the tensiometers and buried columns were located.

Although no numerical data on peg movement were obtained from the transverse peg lines, field observation and photographs indicated that the pegs inserted to 10 cm generally moved more than those inserted to 20 cm, reflecting the decrease in soil movement with depth shown by the buried columns.

THE MECHANICS OF SOIL MOVEMENT

The maximum potential soil creep, calculated on the basis of the experimental results, indicates that potential creep values were often less than the actual recorded soil movement. The total creep movement is likely to be considerably less than the potential value (Washburn, 1967), and it may therefore be concluded that solifluction, in addition to frost creep, is active in causing down-slope soil movement in this area.

The profiles of the buried columns suggest movement by viscous flow, each soil layer moving slightly more per year than that immediately below it, the whole moving mass of soil thereby constituting a shear zone. The fact that the liquid limits of the soils are low indicates that, for the period during spring when saturated conditions prevail, the soil is in a state of potential flow. However, in column 2, site C, there is some evidence of shearing between the uppermost undisturbed column section and the lowermost disturbed section (Fig. 9).

Under the soil conditions during thaw, with saturated surface soil layers above a frozen subsoil, it seems possible for slip to occur along a shear plane immediately above the frozen subsoil, producing a slab slide. Assuming uniform shear strength, groundwater at the surface and a shear-plane parallel to the surface, it is fairly simple to compute the critical angle at which this type of failure is likely.[2] A summary of the results of this stability analysis is given in Table VI. It can be seen that the angles of slope in the study area are generally lower than critical, although the upper slope at A (17°) is steeper than the

TABLE VI

Summary of the slope stability analysis

Location	General slope angle		Soil density kg/m³	Saturated water content % dry weight	Critical slope peak strength	Critical slope residual strength
	(a) upper slope	(b) lower slope				
Site A	17°	5°	2058	24	19° 48'	15° 54'
B	5°	10°	1980	25.3	19° 06'	15° 20'
C	6°	13°	1955	29.4	18° 53'	15° 09'

critical value under residual shear-strength conditions. Thus if failure has occurred in the past, leaving the soil under residual strength conditions, this slope is in a potentially unstable state today. However, since the liquid limit of the soil is reached before total saturation, it seems likely, even on the steepest slope, that soil movement by flow would take place before slip failure. Shearing within column 2, site C, may be the result of locally generated, high pore-water pressures causing local shear, or it may be that the column was disturbed during insertion producing this apparently sheared effect.

DOWNSLOPE SOIL MOVEMENT ELSEWHERE IN OKSTINDAN

Measurement of rates of soil movement elsewhere in Okstindan was not possible, but much may be learnt of the nature of soil movement within similar mass-movement features from a simple mechanical analysis of their soils. Samples were taken from various turf-banked and stone-banked lobes on the north and east flanks of Okstindan (Fig. 1) and their grain size, index properties and loss on ignition were found. These are shown, together with local slope angles, in Table VII.

TABLE VII

Soil properties within solifluction lobes in north-eastern Okstindan

Location	Clay content %	Silt content %	Sand & gravel content %	Liquid limit %	Plastic limit %	Plasticity index %	Loss on ignition %	Slope angle
Site A	8·4	15·2	76·4	22·7	20·3	2·4	2·17	17°/5°
Site B	11·6	13·9	74·5	24·4	21·3	3·1	3·57	10°
Site C	4·7	11·6	83·7	20·7	17·8	2·9	1·53	13°
Sample 1	5·0	4·0	91·0	24·1	18·4	3·7	2·98	15°
Sample 2	7·5	6·5	86·0	30·6	23·7	6·9	2·69	13°
Sample 3	6·0	10·0	84·0	29·0	23·1	5·9	4·00	10°
Sample 4	14·0	8·0	78·0	28·0	26·0	2·0	3·49	12°
Sample 5	9·0	8·0	83·0	33·5	25·0	8·5	4·16	15°
Sample 6	10·0	9·0	81·0	65·0	49·0	16·0	11·80	18°

All the lobes sampled were fed by snowmelt during spring from late-lying snow patches immediately up-slope, and all had very high moisture content when sampled in June 1971. Grain-size distributions within all six lobes were essentially similar to those of the study area, with high sand contents and relatively low clay contents. When plotted and compared with Beskow's (1935) limits, all showed marginal frost susceptibility, suggesting that frost creep contributes to down-slope soil movement within the lobes, its amount probably depending chiefly on moisture supply and depth of snow during freezing.

The sandy nature of the soils suggests high internal angles of friction and this, coupled with their low plasticity and liquid limits, indicates that down-slope movement under saturated conditions probably occurs by viscous flow rather than slip failure. It appears, therefore, that solifluction and soil creep are the chief agents of soil movement within these lobes, soil movement being similar in nature to that described in detail within the study area.

CONCLUSIONS

(1) Frost heave is greatest when penetration of the freezing plane through the soil is slow.
(2) Soil creep results from both frost heave and expansion and contraction of the soil owing to varying moisture content.
(3) Saturated conditions occur for a short period during the spring thaw as a result of drainage being impeded by a frozen subsoil layer.
(4) Maximum annual soil movement occurs where saturated conditions persist longest, not where the slope is greatest.
(5) Movement takes place by a combined process of creep and flow.

This project is necessarily a long-term venture, and it is hoped that further information on the rates and nature of down-slope soil movement within this area of north Norway will be obtained in succeeding years.

ACKNOWLEDGEMENTS

This work was carried out as part of a Ph.D. research programme at the University of Reading, and was supported by an N.E.R.C. research grant. Thanks are due to Dr Peter Worsley and other members of the Reading University Okstindan Research Project for their invaluable assistance with field work. The University College of Swansea kindly contributed to the cost of illustrations.

NOTES

1. A Grant Model D miniature temperature recorder was used in this study.
2. The infinite slope analysis of A. W. Skempton and F. A. DeLory (1957) was used, assuming steady seepage parallel to the ground surface, where F, the factor of safety, can be expressed as

$$F = \left(\frac{\gamma'}{\gamma}\right) \frac{\tan \phi'}{\tan \beta}$$

β is the slope angle and γ and γ' are the saturated and submerged densities respectively.

REFERENCES

ANDERSSON, J. G. (1906) 'Solifluction, a component of sub-aerial denudation', *J. Geol.* 14, 91–112
BENEDICT, J. B. (1970) 'Down-slope soil movement in a Colorado Alpine region: rates, processes and climatic significance', *Arct. alp. Res.* 2, 165–226
BESKOW, G. (1935) 'Tjälbildningen och tjällyftningen med särskild hänsyn till vägar och järnvägar', *Sver. geol. Unders. Afh.*, Ser. C, 375
CASAGRANDE, A. (1932a) 'Research on the Atterberg limits of soils', *Publ. Rds, Wash.* 13, 121–36
HARRIS, C. (1971) 'Preliminary investigations into solifluction processes, Austre Okstindbreendal', Okstindan Research Project 1969, prelim. Rep., Univ. of Reading, 6–14

HIGASHI, A. (1958) 'Experimental study of frost heaving', *Rep. Snow Ice Permafrost Res. Establ.* 45, 44 pp.

JAHN, A. (1961) 'Quantitative analysis of some periglacial processes in Spitsbergen', *Nauka ziemi* 2, 54 pp.

RUDBERG, S. (1958) 'Some observations concerning mass movements on slopes in Sweden', *Geol. För. Stockh. Förh.* 80, 114–25

SKEMPTON, A. W. and F. A. DeLORY (1957) 'Stability of natural slopes in London Clay', *Proc. 4th. int. Conf. Soil Mech.* 2, 378–81

TABER, S. (1930) 'The mechanics of frost heaving', *J. Geol.* 38, 303–17

WASHBURN, A. L. (1967) 'Instrumental observations of mass-wasting in the Mesters Vig district, north-east Greenland', *Meddr Grønland* 166 (4), 297 pp.

WILLIAMS, P. J. (1959) 'An investigation into processes occurring in solifluction', *Am. J. Sci.* 257, 481–90

YOUNG, A. (1960) 'Soil movement by denudation processes on slopes', *Nature, Lond.* 188, 120–2

RÉSUMÉ. *Une mise au point des procès de mouvement du sol dans les gradins végétaux de solifluction, à Okstindan, Norvège du nord.* La région étudiée consiste à une côte exposée à l'est d'entre 4° et 20° à une altitude d'environ 710 m, étant développée sur un dépôt glaciaire sablonneux, et terrassée contre une roche moutonnée. Sur cette côte se sont développés des gradins végétaux de solifluction ayant une hauteur frontale qui varie de 0,5 à 1,5 m et une longueur axiale d'entre 10 m et 50 m. On a mesuré les paramètres suivants: (i) les conditions d'humidité du sol, au moyen de tensiomètres d'humidité de terre et d'échantillonnage hebdomadaire; (ii) la hauteur de soulèvement hivernal du sol grâce à l'action de la gelée à trois endroits, en utilisant des cadres fixes de fer cornier et des tiges mobiles verticales; (iii) le mouvement total annuel du sol superficiel au moyen d'une revue de chevilles de stations fixes de théodolite; (iv) des profils verticaux de mouvement avec profondeur, au moyen de la méthode de la colonne enterrée; (v) les caractéristiques de force du sol, en se servant et d'un appareil trixial et d'une boîte; (vi) les limites d'Atterberg, à des profondeurs de 10, 50 et 100 cm. On traite de la mécanique du mouvement de terre en utilisant les données des sousdites dimensions; et on considère la possibilité de mouvement de terre sur des plans de glissement (définis) par opposition à une sorte d'écoulement de terre.

FIG. 1. Carte de situation

FIG. 2. La région étudiée; la location et l'instrumentation des sites des expériences. Carte ajoutée, des plans de profil des sites

FIG. 3. Partie de la région étudiée. Hauteur des marches à peu près 50 cm

FIG. 4. Composition et dimensions des grains à deux endroits de chaque site; des échantillons creusés à des profondeurs de 10, 50, et 100 cm, au-dessous de la surface

FIG. 5. Températures de la terre au printemps, sites A et B, 1970

FIG. 6. Températures de la terre en automne, sites A et B, 1970

FIG. 7. Chiffres de tensiometers, été 1970. On donne ceux-ci dans des divisions d'échelle et les parties noicies qui se trouvent au-dessous des courbes représentent les périodes pendant lesquelles la terre est saturée.

FIG. 8. Mouvement de la terre, 1970–71

FIG. 9. Mouvement de la terre à un niveau profond, 1969–70

ZUSAMMENFASSUNG. *Eine Untersuchung der Erdbewegungsvorgänge in den Torfstauungssolifluktionslappen in Okstindan, Nordnorwegen.* Das Untersuchungsgebiet besteht aus einem gegen Osten gewandten Abhang, der zwischen 4° und 20°, in einer Höhe von 710 m liegt. Der Abhang hat sich auf sehr feinem Gletschersand (sandy silt till) entwickelt, und hat sich gegen eine roche moutonnée gehäuft. Auf diesem Abhang haben sich Torfstauungssolifluktionslappen entwickelt, deren frontale Höhen zwischen 0,5 bis 1,5 m und Achsenlängen zwischen 10 bis 50 m liegen. Die folgenden Parameter wurden gemessen: (i) der Erdfeuchtigkeitszustand mit einem Erdfeuchtigkeitstensiometer und wöchentlichen Proben; (ii) der Betrag der Winterfrosthebung an drei Stellen durch feststehende Winkeleisenrahmen und senkrecht bewegende Stangen; (iii) die gesamte jährliche Erdoberflächenbewegung durch Wiedervermessung der Dübel von feststehenden Theodolitstationen; (iv) senkrechte Profile der Bewegung mit Tiefe mittels der vergrabenen Säulenmethode; (v) Scherenstärkecharakteristiken der Erde, wo man den triangulären Scherenapparat und eine Scherendose benutzt; (vi) Atterberg Grenzen in 10, 50 und 100 cm Tiefe. In dem man die Angabe der oben genannten Vermessungen benutzt, wird die Mechanik der Erdbewegung behandelt und die Möglichkeit des Erdrutschens entlang der bestimmten Scherenflächen, im Gegensatz zu einer stromartigen Erdbewegung, erwägt.

ABB. 1. Lagekarte

ABB. 2. Das Untersuchungsgebiete und deren Instrumentierungen dar. Im Eckeneinsatz: Abhangsprofile der Experimentgebiete

ABB. 3. Tiel des Untersuchungsgebietes, erhöhungen circa 50 cm

ABB. 4. Komposition der granulären Grösse; Proben wurden aus 10, 50, und 100 cm tiefe unter der Oberfläche aus stellen in jedem Gebiet entnommen

ABB. 5. Fruhlingserdtemperaturen, Stellen A und B, 1970

ABB. 6. Herbsterdtemperaturen, Stellen A und B, 1970
ABB. 7. Tensiometerlesungen, Sommer 1970. Die Lesungen sind in Abstufungseinteilungen dargestellt und die schattierten Gebiete unter den Graphen stellen Zeiträume der Erdsaturation dar
ABB. 8. Erdbewegung 1970–71
ABB. 9. Erdbewegung mit Tiefe, 1969–70

The nature of the ice-foot on the beaches of Radstock Bay, south-west Devon Island, N.W.T., Canada in the spring and summer of 1970

S.B.McCANN

Associate Professor of Geography, McMaster University

AND R.J.CARLISLE

Graduate student in Geography, McMaster University

Revised MS received 10 November 1971

ABSTRACT. The ice-foot present along the coast of south-west Devon Island in the early summer of 1970 is described and a series of profiles across the feature, surveyed at intervals during the period of break-up, are presented. The pattern of break-up for both the sea ice and shore-fast ice is documented. Data on the relationship between (a) beach slope and (b) ice-foot thickness and width are derived from the profiles, and indicate that increases in the former result in decreases in the latter.

THE ice-foot—'that part of the sea ice which is frozen to the shore and is therefore unaffected by tidal movements' (R. Bentham, 1937, p. 328)—was a pronounced feature of the shore zone throughout much of the southern Queen Elizabeth Islands, in the Canadian Arctic, during the late spring and early summer of 1970. This paper describes the form of the ice-foot and its decay, based on repeated observations and surveys made from mid-June onwards at a study site on Radstock Bay, south-west Devon Island (Fig. 1). Certain problems of definition are discussed and the conditions likely to lead to a well-developed ice-foot are considered. The relations between the form, that is the thickness and width, of the ice-foot on Radstock Bay, and beach slope at different localities are examined in an attempt to isolate the effect of beach geometry on ice-foot development.

The research is part of a longer-term study of the form and processes of the modern beach zone in the high Arctic which was initiated in 1968 at sites on south-west Devon Island, following a reconnaissance in 1967 in the Resolute area of Cornwallis Island. Some of the results, concerning the general role of ice, beach sediments and permafrost, have already been published (S. B. McCann and E. H. Owens, 1969; 1970; Owens and McCann, 1970; McCann and F. G. Hannell, 1971) and these papers provide background information on the study area. Certain basic facts about the beach régime are, however, relevant in this introduction. The tides at Radstock Bay are of the mixed semi-diurnal type with a spring range of 2·5 m, and the period of potential wave action on the beaches is limited to 8–10 weeks in the year when the beaches are free of ice and there is open water to the south.

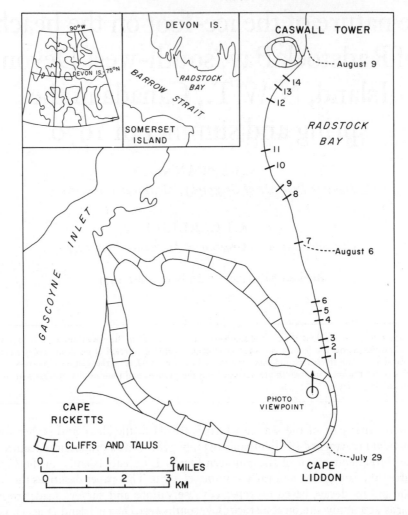

FIGURE I. Location map of the study area, showing the profile sites and the positions of the ice margin in Radstock Bay in late July–early August, 1970

The sea-ice cover in Barrow Strait, south of Radstock Bay, generally breaks up in late July and re-forms sometime in late September–early October, though there is considerable variation from year to year in the dates of both disintegration and re-formation. Large amounts of pack ice may remain in Barrow Strait during the 'open water' period, inhibiting wave generation, and 'fast' ice may remain on the beaches throughout much of the time, preventing waves from attacking the beach sediments. The beaches are thus low-energy, meso-tidal in régime and they are built very largely of material of gravel size. Cliffs are an important part of the coastal landscape, but the study beach, which stretches for 8 km along the outer western shore of Radstock Bay, between Cape Liddon and Caswall Tower, is backed by a fine series of raised marine strandlines (Fig. 2A and 3A).

The term ice-foot has a long usage in the literature on Arctic and Antarctic coasts and

FIGURE 2A. General view of the outer part of Radstock Bay on 16 July, with complete ice cover, taken from the viewpoint indicated on Figure 1. B. The shore on 16 July, showing the ice-foot, the series of free-floating floes in the tidal leads, and the sea ice proper

FIGURE 3A. General view of the outer part of Radstock Bay on 6 August, taken from the viewpoint indicated on Figure 1. The mass of ice in the right foreground is moving out of the bay, leaving a fringe of beach-fast ice in the form of an ice-foot along the shore. B. The shore on 7 August, showing the ice-foot and in the background the ice margin across the bay as it remained from 6–9 August some of the more important references are reviewed by R. W. Feyling-Hanssen (1953), though the early discussion and classification of the phenomena by C. S. Wright and R. E. Priestley (1922), in their classic work on Antarctic glaciology, is overlooked. The definition by Bentham (1937) quoted in part at the beginning of this paper and the similar one, based on Wright and Priestley's account, by J. R. F. Joyce (1950)—'the ice formation which joins the land and sea ice between high- and low-water marks'—indicate an important genetic point in that the development of an ice-foot has been linked to a tidal environment. Indeed Bentham states that one of the conditions necessary for the formation of an ice-foot is a considerable tidal range. The term ice-foot is widely used in Greenland in the same sense (G. H. Petersen, 1962). In such an environment, when temperatures fall low enough in the autumn, the beaches become covered with a layer of ice as the tide recedes to the ebb: the process is repeated with each tide until a continuous rim of ice develops along the coast, which builds out seawards from high-water mark. Storm-wave conditions at the time of formation would result in the accumulation of ice above the normal high-water mark, by the freezing of swash and spray, and in the formation of a storm ice-foot (Wright and Priestley, 1922). Other variations in form might be introduced if grounded pack ice is incorporated in the ice-foot or if ice-foot buckles owing to the pressure of pack ice driving onshore under high winds, but the essential control of ice-foot width and thickness remains the same—the tidal rise and fall allows fast ice to develop in a broad thick, band along the shore. The slope and, correspondingly, the width of the beach in the intertidal zone are additional controlling factors. The term ice-foot has, however, been used to describe the beach-fast ice in non-tidal areas such as the Great Lakes (J. H. Zumberge and J. T. Wilson, 1954) and in microtidal areas such as northern Alaska (R. W. Rex, 1964). Rex describing the beaches of Barrow, Alaska, where the normal tidal range is of the order of only 0·15 m, reports particularly on the development of a storm ice-foot, which he suggests might better be called the gravel-sand-ice foot, because of the inclusion of large amounts of beach sediment. The term 'kaimoo' is used by G. W. Moore (1966) to describe a feature, which seems essentially the same as Rex's gravel-sand-ice foot, developed in a similar microtidal environment in north-west Alaska. This is the flat-topped ice-and-gravel rampart, commonly up to 1·5 m thick, formed on the surface of the beaches as a result of the freezing of wave swash. The use of the term is perpetuated, with an explanatory diagram, in Moore's account of Arctic beaches in the *Encyclopaedia of geomorphology* (1968) but there is no reference to the older and possibly more inclusive term, ice-foot. More recently, H. G. Green (1970), in an account of beach features in the microtidal (0·3 m range) environment near Nome, Alaska, defines the kaimoo as the bed of ice and frozen sand and gravel which extends shoreward from the waterline, and the ice-foot as the fringe of ice extending seawards from the waterline. Moore (1968, Fig. 1) is content to label this latter feature simply as 'fast ice'. There is thus some confusion in terminology which may be resolved if the terms kaimoo and ice-foot are restricted to forms developed under different tidal conditions. The kaimoo is strictly a result of swash action in a non-tidal or microtidal environment: it is built up above still water level and usually contains interbedded sediment, washed up from the water-covered beach below. The ice-foot is developed in the upper part of the intertidal zone of tidal beaches: it may be built considerably above high-water level by the freezing of swash and spray, and clearly extends well below high-water mark. It is unlikely to contain much interbedded sediment, for the lower part of the beach below the ice-foot is likely to be sealed by a cover of ice developed at low tides.

DISINTEGRATION SEQUENCE AND EMERGENCE OF THE ICE-FOOT

Observations were commenced on 18 June when the field party arrived at the study site. At this time a complete cover of sea ice up to 2 m thick existed in Radstock Bay and in Barrow Strait to the south. There was an almost complete snow cover on the land surface and deep drifts were present along the shore, so that it was impossible to determine the exact position of the shoreline or the form of the beach-fast ice. However, it was possible, with care, to locate the series of narrow cracks in the snow cover, which indicated the position of the tidal cracks below, and so fix the outer limit of the fast ice. Melting of the snow cover was already underway at this time and during the period 21–26 June large puddles developed on the surface of the sea ice in the bay, some of the meltwater draining into the tidal cracks and thus widening them. At this time, too, snow lying on the backshore zone of the beach began to melt and drain seawards to the tidal cracks in a series of small meltwater channels. As melting proceeded more rapidly in the last week of June, water became ponded at the rear of the beach and channels, up to 2 m deep, were eroded through the packed snow and beach-fast ice towards the tidal cracks, which by 1 July were wide enough at certain stages of the tide to be more appropriately termed 'tidal leads'. By the beginning of July also it became clear that the beach-fast ice took the form of a pronounced ice-foot which had been built up the previous autumn above the level of normal high water.

The action of running water in the beach zone was very marked in the spring of 1970 owing to the greater amount of meltwater resulting from a heavier snow cover than normal, but it may be observed every year. In addition to its importance in the local sorting of beach sediments and in producing small-scale depositional features, such as micro-deltas, it clearly plays an important role in the initial breaking up of the fast ice. Meltwater channels cutting across the ice-foot occurred every 20–30 m along the beach at Radstock Bay.

Meltwater had ceased to flow on the beach by 8 July and there was no significant change until 13 July when it was observed that the sea ice separating the widening tidal leads near Caswall Tower was beginning to break up into a series of free-floating ice floes. Figure 2B shows this condition on 16 July at a point midway along the beach. During the next 10 days the process of melting and disintegration in the tidal crack zone proceeded rapidly, so that by 23 July the situation was one of a single broad tidal lead all along the shore, with a series of small ice floes contained within it. The sea ice in Radstock Bay and the fast ice along the beach, in the form of an ice-foot, were thus quite separate by this time, and conditions within the bay would have permitted the ice to move out, for certain major leads crossing the bay were also fully open by this time. However, this did not occur for another two weeks, for a considerable ice cover continued to exist in Barrow Strait to the south. The ice at the·entrance to Radstock Bay did not break and move out until 29 July and it was not until 6 August that suitable wind conditions existed to begin to move the ice out of the outer part of the bay itself, a process completed by 9 August. Figure 3A illustrates how the ice moves out of the bay in very large masses: the moving ice in the photograph is that which moved out on 6 August (see also Figure 1 for various positions of the ice margin in the bay). The photograph also shows clearly the ice-foot remaining along the shore, and Figure 3B, taken on the beach the next day, shows the nature and scale of this feature. By 15 August, that is 8 days after the photograph was taken, the ice-foot had gone completely. During the melting period it was possible to ascertain that the ice-foot had undergone very little deformation by buckling and that it contained very little beach sediment.

THE FORM OF THE ICE-FOOT

The form of the ice-foot was measured at a series of fourteen beach profile sites along the outer western arm of Radstock Bay (Fig. 1) on various occasions during the ablation and break-up period. Profiles were recorded at all sites on 8 July, 13 July and 15 August, and at certain sites on 24 June, 2–3 July, 24 July and 8 August. Figure 4 shows three selected profiles for each site superimposed on the same diagram and indicates the general form and progress of decay of the ice-foot. With the exception of sites 1–3, surveyed on 13 July, the first profile shown was measured on 2–3 July, when the snow lying on the ice had melted and the ice-foot had emerged as a distinct feature. These profiles are continued seawards across the zone of tidal cracks to the sea ice proper. The second set of profiles measured for sites 4–14 on 8 August and for sites 1–3 on 12 August, indicates the condition of the ice-foot after considerable ablation had taken place, and the last set of profiles records the slope of the beach after the final decay of the ice, but before any significant wave action had taken place.

There was an overall consistency of form all along the beach, in which the most striking feature was the steep seaward edge of the ice-foot early in the season. This near-vertical face along the line of the most shoreward tidal crack was commonly of the order of 2·5–3 m high. The highest point on the ice was frequently within 5 m of this first tidal crack and there was usually a slope inwards from this point towards the beach. The mean width of the ice-foot at the fourteen sites was 16·2 m, within a range from 9·3 to 21·5 m, but on ten of the profiles it was between 15 and 19 m. The mean maximum thickness of the ice-foot at the fourteen sites was 2·2 m, within a range from 1·4 to 2·9 m, with nine profiles showing thicknesses between 2 and 2·5 m. There is no noticeable trend in the values of either thickness or width along the beach between profile 1 and profile 14.

The position of the ice-foot in relation to tidal levels is illustrated in Figure 5, in which the vertical extent of the formation at each site is plotted as a vertical line on a set of horizontal axes representing the various mean tide levels of the mixed semi-diurnal tide. The spacing of the profiles on this diagram accords with their relative positions in the field. The top of the ice-foot in each case is above lower high-tide level and in all but one case is above mean higher high-tide level. In three instances, at sites 7, 8 and 9, the top of the ice-foot is a metre or more above the level of higher high water. This would seem to indicate that the ice-foot was developed under fairly open water conditions with quite large waves, for it would require the freezing of wave swash and spray to raise the level of the feature above the normal still-water high-tide level. The bottom of the ice-foot is more consistent than the top, and is in most cases at about the level of the mean higher low tide. Thus the feature present in Radstock Bay in the spring and early summer of 1970 almost completely covered the intertidal zone of the beaches.

Figure 6 illustrates the relationships which exist between beach slope and ice-foot thickness and width at the fourteen sites along the beach. Beach slope was measured between the positions of lower high tide and higher low tide and is expressed in degrees on the diagrams. Slopes ranged from 5·9° to 10·7°, with a mean slope of 8·2°, and there is no regular variation in slope along the beach. The two sets of data plotted on Figure 6 show that a clear relationship exists between the slope of the beach in the intertidal zone and the thickness and width of the ice-foot as present in Radstock Bay in 1970. As beach slope increases, both the thickness and width of the ice-foot tend to decrease.

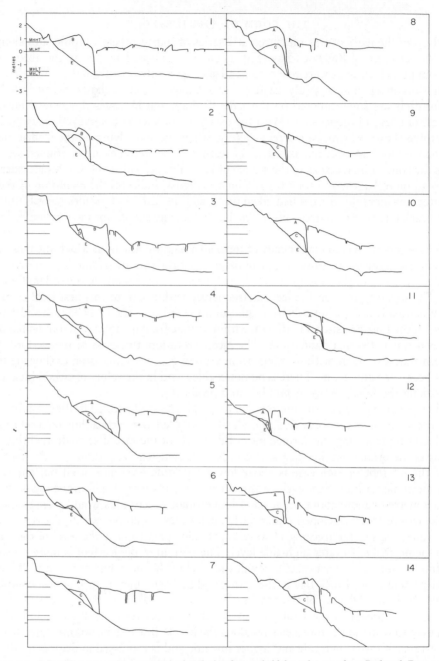

FIGURE 4. Selected cross-sections across the beach, ice-foot and tidal crack zone along Radstock Bay, surveyed in July and August, 1970.
(A—2-3 July; B—13 July; C—8 August; D—12 August; E—15 August). Profile locations are shown on Figure 1. Vertical exaggeration is × 5. MHHT, MLHT—mean higher and mean lower high tide, respectively

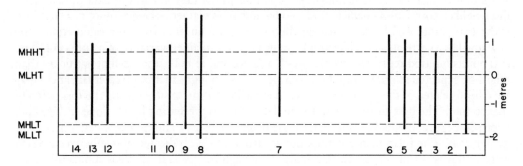

FIGURE 5. The position of the ice-foot in relation to mean tide levels (abbreviations as in Figure 4)

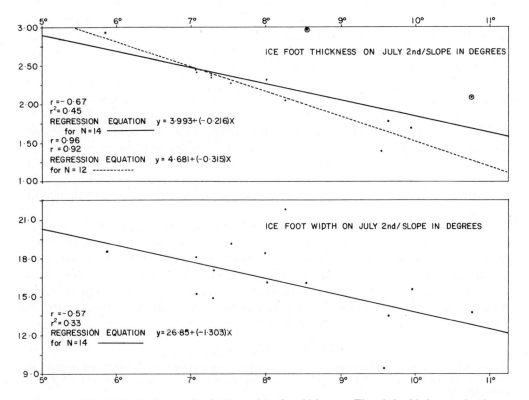

FIGURE 6A. The relationship between beach slope and ice-foot thickness. B. The relationship between beach slope and ice-foot width

Two regression lines have been fitted to the points on Figure 6A, which shows beach slope plotted against ice-foot thickness. The line which best fits all the data points (solid line) yields a correlation coefficient of -0.67 and an explained variance of 45 per cent but, if the two points which are ringed on the diagram are omitted from the calculations, the new line has a correlation coefficient of -0.96 and an explained variance of 92 per cent. The two points omitted represent profile sites 8 and 10, and at both, the ice-foot is thicker than might be expected in relation to the beach slope. It is thought that the greater thickness may have been related to the presence of grounded ice floes at these sites during the formation of the ice-foot. The best-fit regression to the data points representing beach slope and ice-foot width in Figure 6B yields a correlation coefficient of -0.57 and an explained variance of 33 per cent. The relationship is thus not as strong as that indicated above between beach slope and ice-foot thickness. In this analysis the value of r was in all cases sufficiently high for it to be significantly different from zero at the 0.99 per cent confidence level.

CONCLUSION

The form of the ice-foot along Radstock Bay in the spring and summer of 1970 was the most pronounced observed in three seasons' field work on the coast of south-west Devon Island. Similar features had been present in the previous two years but they did not extend so high above normal high-tide level and did not last so long into the break-up period. The size and extent of the ice-foot in any year clearly depend on the sea conditions existing at the time of freezing in the previous autumn and it is thought that the 1970 ice-foot originated in conditions of medium-sized waves for the area. As there were only a few floes included in the ice-foot and little evidence of buckling it is concluded that the bay was relatively ice-free at the time of ice-foot formation, probably in mid- or late October 1969.

The importance of tidal range in ice-foot development has been considered in the introduction to this paper. The range of 2·5 m in Radstock Bay facilitated the development of an ice-foot some 15–19 m wide in the autumn of 1969, which virtually covered the inter-tidal zone of the beach and rose to heights in excess of a metre above mean high-tide level. The control of beach slope on ice-foot width and thickness, as seen in the Radstock Bay measurements, is such that as slope increases so the width and thickness of the ice-foot decrease.

The disintegration sequence of both fast ice and sea ice described for Radstock Bay in 1970 can be considered as fairly typical for this area. The development of the tidal crack into a single tidal or shore lead is important in isolating the ice-foot from the sea ice proper; and the erosion of channels by water draining seawards from the melting snow in the backshore zone, early in the season, is important in breaching the ice-foot. The sea ice had moved out of the outer part of Radstock Bay, adjacent to the study beach, by 9 August, and the ice-foot had been ablated and eroded away by 15 August.

ACKNOWLEDGEMENTS

The research for this paper was supported in part by the Defence Research Board of Canada (Grant number 9511–106) and by Grants in Aid of Research from the National Research Council of Canada (Grant number A5082) and McMaster University.

REFERENCES

BENTHAM, R. (1937) 'The ice-foot', Appendix III in *Arctic Journeys. The story of the Oxford University Ellesmere Land Expedition* (E. Shackleton)

FEYLING-HANSSEN, R. W. (1953) 'Brief account of the ice-foot', *Norsk geogr. Tidsskr.* 14, 45–52

GREENE, H. G. (1970) 'Microrelief of an arctic beach', *J. sedim. Petrol.* 40, 419–27

JOYCE, J. R. F. (1950) 'Notes on ice-foot development, Neny Fjord, Graham Land, Antarctica', *J. Geol.* 58, 646–9

McCANN, S. B. and F. G. HANNELL (1971) 'Depth of the "frost table" on Arctic beaches, Cornwallis and Devon Islands, N.W.T., Canada', *J. Glaciol.* 10, 155–7

McCANN, S. B. and E. H. OWENS (1969) 'The size and shape of sediments in three Arctic beaches, S.W. Devon Island, N.W.T., Canada', *Arct. alp. Res.* 1, 267–78

McCANN, S. B. and E. H. OWENS (1970) 'Plan and profile characteristics of beaches in the Canadian Archipelago', *Shore Beach* 38, 26–30

MOORE, G. W. (1966) 'Arctic beach sedimentation' in *Environment of the Cape Thompson region* (ed. N. J. WILIMOVSKY and J. N. WOLFE), 587–608

MOORE, G. W. (1968) 'Arctic beaches' in *Encyclopaedia of geomorphology* (ed. R. W. FAIRBRIDGE), 21–2

OWENS, E. H. and S. B. McCANN (1970) 'The role of ice in the Arctic beach environment with special reference to Cape Ricketts, S.W. Devon Island, N.W.T., Canada', *Am. J. Sci.* 268, 397–414

PETERSEN, G. H. (1962) 'The distribution of *Balanus balanoides* (*L.*), and *Littorina Saxatilis, Olivi*, var. Groendlandica, Mencke in Northern West Greenland', *Meddr Grønland* 159, 9, 36–40

REX, R. W. (1964) 'Arctic beaches, Barrow, Alaska' in *Papers in marine geology* (ed. R. L. MILLER), 384–400

WRIGHT, C. S. and R. E. PRIESTLEY (1922) *British (Terra Nova) Antarctic Expedition*

ZUMBERGE, J. H. and J. T. WILSON (1954) 'Effect of ice on shore development', *Proc. 4th Conf. Coastal Engng*, 201–5

RÉSUMÉ. *Les caractéristiques du « Ice Foot » sur les plages de Radstock Bay, s-o. de l'Île Devon, T.N.O., Canada, dans le printemps et l'été de 1970.* Cet article donne une description du « ice foot » marqué qui fut observé le long de la côte du sud-ouest de l'Île Devon dans le printemps et l'été de 1970. Il présente aussi une série de profils traversant le phéno-mène, desquels on a fait le relevé de temps en temps dans la période de désintégration. On dépose la succession d'évènements durant cette période, pour la glace de la mer et aussi pour la glace attachée à la plage. L'analyse des pro-fils donne des valeurs pour le rapport entre la pente de la plage et l'épaisseur et la largeur du « ice foot »; ces valeurs indiquent que, plus la pente de la plage est raide, le « ice foot » est moins épais et moins large.

FIG. 1. Carte de la localité étudiée, démontrant les sites des profils, et les positions du bord de la glace dans Radstock Bay dans la seconde moitié de juillet et la première moitié d'août, 1970

FIG. 2A. Vue générale de la bouche de Radstock Bay, le 16 juillet; couverture de glace totale; photographie faite de la pointe indiquée à la Fig. 1. B. Vue de la plage, le 16 juillet, démontrant le « ice foot », la série de banquises détachées dans les fissures produites par la marée, et la vraie glace de la mer

FIG. 3A. Vue générale de la bouche de Radstock Bay, le 6 août; photographie faite de la pointe indiquée à la Fig. 1. Le bloc de glace à droite dans le premier plan quitte la baie, laissant rester une bordure de glace attachée à la plage sous forme d'un « ice foot ». B. Vue de la plage, le 7 août, démontrant le « ice foot » et, dans l'arrière-plan, le bord de la glace à travers de la baie, comme elle est restée après le 6–9 août

FIG. 4. Profils choisis, à travers de la plage, du « ice foot » et de la zone des fissures produits par la marée, le long de Radstock Bay; relevés faits en juillet et en août, 1970. (A—le 2–3 juillet; B—le 13 juillet; C—le 8 août; D—le 12 août; E—le 15 août). Les sites des profils sont données à la Fig. 1. L'exagération de l'axe verticale est de × 5. MHHT—niveau moyen supérieur de la marée haute; MLHT—niveau moyen inférieur de la marée haute

FIG. 5. Les positions du « ice foot » en rapport avec les niveaux moyens de la marée (abréviations comme pour la Fig. 4)

FIG. 6A. Le rapport entre la pente de la plage et l'épaisseur du « ice foot ». B. Le rapport entre la pente de la plage et la largeur du « ice foot »

ZUSAMMENFASSUNG. *Die Natur des ,Ice-foot' an den Stränden von Radstock Bay, S.W. Devon Insel, N.W.G., Kanada im Frühling und Sommer 1970.* Der ausgesprochene ,ice-foot' entlang der Küste von S.W. Devon Insel Anfang des Sommers 1970 wird beschrieben, und eine Serie von Profilen über die Eigenschaft, in Abständen während der Bruchperiode vermessen, wird gezeigt. Das Bruchmuster, sowohl des Meereseises, als auch des Uferschnelleises, wird dokumentiert. Angaben über die Beziehung der Strandneigung und der ,ice-foot' Dicke und Weite werden von den Profilen abgeleitet und zeigen an, dass Zunahmen in jener in Abnahmen der letzteren zwei resultieren.

ABB. 1. Lagekarte des Studiengebietes, die die Profillagen und die Positionen des Eisrandes in der Radstock Bucht Ende Juli—Anfang August 1970 zeigt

ABB. 2A. Allgemeine Betrachtung des äusseren Teiles der Radstock Bucht am 16. Juli, mit kompletter Eisdecke, vom Blickpunkt, angegeben in Abb. 1, genommen. B. Das Ufer am 16. Juli, das den ,ice-foot', die Serie von freischwim-menden Eisfeldern in den Gezeiten-Führungen und das eigentliche Meereseis zeigt

ABB. 3A. Allgemeine Betrachtung des äusseren Teiles der Radstock Bucht am 6. August, vom Blickpunkt angegeben, in Abb. 1, genommen. Die Eismasse im rechten Vordergrund bewegt sich aus der Bucht, einen Rand von Strandschnel-leis in der Form eines ,ice-foot' entlang des Ufers zurücklassend. B. Das Ufer am 7. August, zeigt den ,ice-foot' und im Hintergrund den Eisrand über der Bucht wie es vom 6–9 August zurückblieb

ABB. 4. Ausgewählte Querschnitte über den Strand, ,ice-foot' und Gezeiten-Splitterzone entlang der Radstock Bucht, gemessen im Juli und August 1970. (A—Juli 2–3; B—Juli 13; C—August 8; D—August 12; E—August 15). Profillagen werden in Abb. 1 gezeigt. Senkrechte Übertreibung ist durchschnittlich höhere und durchschnittlich tiefere Flut, entsprechend: MHLT, MLLT = durchschnittlich höhere und durchschnittlich tiefere Ebbe, entsprechend

ABB. 5. Die Position des ,ice-foot' in Beziehung zu durchschnittlichen Gezeiten-Ständen. (Abkürzungen wie in Abb. 4)

ABB. 6A. Die Beziehung von Strandneigung und ,ice-foot' Dicke. B. Die Beziehung von Strandneigung und ,ice-foot' Weite

The solution of limestone in an Arctic environment

D.I. SMITH

Lecturer in Geography, University of Bristol

Revised MS received 10 November 1971

ABSTRACT. Few detailed observations have been undertaken in high-latitude Arctic environments to establish the form and rate of limestone solution. The solution of limestone may differ from that in lower latitudes and the following conflicting hypotheses are presented in the literature. (1) Enhanced solution in the Arctic may result from colder water temperatures and therefore increased solubility of carbon dioxide, or from particularly high concentrations of carbon dioxide in the air at the bases of snow banks. (2) Reduced solution could be related to the paucity of the soil and vegetation cover.

The study area was situated in a limestone region of north-western Somerset Island at latitude 74° N in Arctic Canada. Some 200 water samples were analysed in the field for their calcium and magnesium content together with limited studies related to pH; additional analyses were undertaken for bedrock samples. The total hardness values were generally less than 95 p.p.m.; taking precipitation and evapotranspiration figures into account, this suggests a weathering rate equivalent to some 2 mm/1000 years. Both the concentration of solutes and the weathering rate are considerably less than those found in lower latitudes and the lack of a close soil cover is thought to be the significant factor. There is evidence to suggest that the limestone solution is concentrated at the snow/rock interface. There is no need to invoke a hypothesis involving a concentration of carbon dioxide greater than that found in the free atmosphere. The area has been ice-free for about the last 10 000 years and the rates of solution indicate that no major development of sub-nival hollows by solutional erosion would have occurred during this period.

NUMEROUS detailed studies concerning the rate of solution of limestone have been published in the last decade. There is, however, a paucity of such work from high-latitude environments. In the absence of detailed studies, most workers in this field have followed the suggestions of J. Corbel (1959a, 1959b) that the solution of limestone in polar regions is greater than in regions of comparable run-off in lower latitudes. Corbel based his statements on the progressive increase in the solubility of carbon dioxide in water with decreasing temperature. Water saturated with carbon dioxide at 0°C and in contact with a continuous supply of air from the normal atmosphere contains almost exactly twice the amount of carbon dioxide of water saturated at 20°C. The carbon dioxide content of the normal atmosphere is slightly in excess of 0·03 per cent by volume. A study by B. Bolin and C. D. Keeling (1965) showed that, for the purposes of this present study, variations of the carbon dioxide content of the atmosphere either in terms of latitude or altitude can be ignored. A carbon dioxide content of 0·03 per cent gives a calcium carbonate solubility at 0°C of some 95 parts per million calcium carbonate (hereafter p.p.m. CaCO₃). Seen in isolation these facts give an oversimplified account of the field situation. C. S. Adams and A. C. Swinnerton (1935) drew attention to the apparent anomaly that the calcium content of natural waters in low latitudes frequently exceeds the calcium saturation values based upon a supply of carbon dioxide from atmospheric sources alone. They suggested that the higher calcium values were possibly related to the percolation of water through the soil, the interstitial air of which had a carbon dioxide content well in excess of that of the air of the normal atmosphere. Subsequent work in temperate and tropical regions has verified this. Although there is still scope for further study, most workers consider that the calcium, and also the magnesium, values are dominantly related to the carbon dioxide content of the soil atmosphere and that the direct action of various organic acids found in the soil is important only in a subsidiary role. The saturated values for calcium and magnesium carbonates found in extra-polar areas exhibit variations but are generally well in excess of 200 p.p.m. As more results become available it appears that variations in the total hardness of water (the sum of p.p.m. $CaCO_3$ and p.p.m. $MgCO_3$) between major climatic regions outside the high latitudes are relatively small but that local variations caused by lithological differences can outweigh such climatic effects.

These facts alone indicate that, in an extreme high-latitude environment for a pure limestone, containing no magnesium, and with no vegetation or soil cover, the calcium saturation value would be some 90 p.p.m. $CaCO_3$. A possible modification of this view arises from the work of J. E. Williams (1949) in the Snoqualmie Pass in the state of Washington. Williams obtained a sample of air from a depth of 2·5 m in a snowbank with a total depth of between 3·0 and 3·6 m. The carbon dioxide content of this sample was 0·098 per cent, clearly well in excess of the normal value obtained from the free atmosphere. In addition, Williams analysed two water samples for their calcium content, the first sample being from the meltwater present in the soil at the base of the snow bank and the second from the Snoqualmie River which drains the region. The calcium content of both these samples was close to 10 p.p.m. $CaCO_3$ and magnesium was present only as a trace. It is unfortunate that no comment is given as to the nature of the bedrock underlying the snow bank or to the regional nature of the lithology of the Snoqualmie River basin.

Subsequent workers have drawn extensively upon the results of Williams particularly quoting the conclusion to his paper which states '. . . chemical weathering, especially that caused by the presence of carbonic acid, can take place under a snow cover. Such weathering will accomplish the greatest results under the deepest drifts, where the snow remains for the longest time, and it must, therefore, contribute to the formation of nivation depressions under the drift' (J. E. Williams, 1949, p. 135).

The explanation offered by Williams for the enriched carbon dioxide content of the atmosphere within the snow is related to limited freeze-thaw action within the snow bank. The re-freezing of limited quantities of melted snow to form ice causes the carbon dioxide content to increase, as the solubility of carbon dioxide in ice is less than for water. Thus on freezing the carbon dioxide content of the interstitial air is increased and subsequent meltwater or rain, passing through the snow bank, absorbs this carbon dioxide to the point of saturation. It should be stressed that the possible enrichment of carbon dioxide from the melting of snow banks is therefore a fundamentally different process from that whereby carbon dioxide is absorbed in meltwater from ice bodies such as glaciers. The content of carbon dioxide for glacial meltwater will be less than the atmospheric content, while that from snow banks, owing to the considerable intergranular pore space, may be greater. Field work in the French Alps and associated laboratory studies by C.Ek (1964) and Ek and A. Pissart (1965) clearly establish the lower carbon dioxide content from glacier meltwater. A study by P. Clement and J. Vaudour (1967) from the southern French Alps considers the pH values of melting snow. The pH of the snow is dominantly related to the content of carbonic acid which is itself related to the carbon dioxide of the snow body, i.e. the ice crystals plus the interstitial air. They find variations with snows of differing ages and types, the newly fallen snow being the most acid with a median value of about pH 5·4 while snows changed by freeze-thaw action have a median value nearer to 6·0. This appears to indicate a reduction of the included carbon dioxide. However, field problems may complicate this apparently simple picture. These include variations in the carbon dioxide content of the snow owing to variations in porosity and the fact that snow samples were only collected from the surface of the snow where the mixing of gases from the snow with the free atmosphere may have taken place. F. D. Miotke (1968) presents a review of the literature concerned with the solubility of limestone in respect of rainwater and water from snow-melt. He also outlines the results of field experiments undertaken in the Picos de Europa region of northern Spain and concludes that water from snow-melt has less potential for dissolving limestone than rainwater and that the rate of solution is slower.

The author is not aware of any detailed observations of this type from high-latitude snow banks. The most pertinent observations would concern the actual carbon dioxide content of the atmosphere of the voids near to the base of the snow bank. Improvements in equipment for the measurement of the carbon dioxide content of gaseous mixtures in the field demonstrate that this is now a feasible proposition (F. Delacour, F. Weissen and Ek, 1968; Miotke, 1968).

Care should be exercised in extrapolating snow-bank observations from high-altitude temperate latitude locations to polar environments for the following reasons. First, it is likely that high-latitude snow banks undergo considerably less freeze-thaw action of the snow which would reduce the effectiveness of the mechanism suggested by Williams; further, the snow types could be expected to be different especially as regards porosity. Secondly, it is possible that in the alpine environments of low latitudes the snow banks overlie patches of soil and vegetation. These are themselves rich in carbon dioxide produced by biogenic processes and, when snow covered, allow diffusion of gases into the overlying snow bank. This, for example, could possibly be the case with the work of Williams.

It is against this background that the observations in Somerset Island were undertaken. The major aspect of the work was to measure the calcium and magnesium content of the water from varying sites, giving particular emphasis to water draining from snow banks. Limited observations were also made regarding the pH of the water draining from the snow banks as this could possibly yield some further information, albeit indirectly, regarding the possibility of an enhanced carbon dioxide content. The primary object, however, was to establish figures for the rate of solutional weathering. The only comparable published figures are from the work of Corbel and these appear to consist of single samples from each locality studied. The figures given by Corbel only relate to calcium content and no magnesium figures are given. Before presenting the results of the work, a brief description of the geology, physiography and climate of the study area is necessary.

DESCRIPTION OF THE STUDY AREA

The area of study was confined to north-western Somerset Island; localities mentioned in the text are shown in Figure 1.

Geology

The area is underlain by Pre-Cambrian and Palaeozoic strata. The actual study area consists of two lower formations, the Aston and Hunting Formations, conformably overlain by Lower Palaeozoic strata divided into the Allen Bay Formation, the Read Bay Formation and

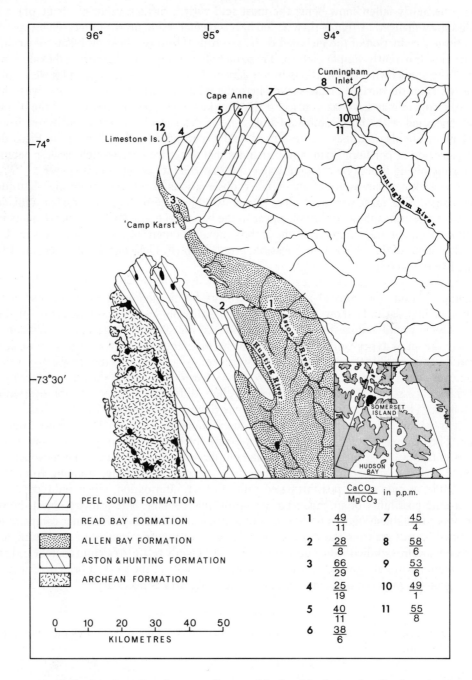

FIGURE 1. Bedrock geology of north-western Somerset Island and hardness values for the major rivers

the Peel Sound Formation. The Aston Formation with a maximum thickness of 800 m has a relatively small outcrop and includes only limited thickness of carbonate rocks. The overlying Hunting Formation reaches a thickness of 2250 m and is dominantly composed of dolostone with minor bands of shale. The Aston and Hunting Formations have been described by M. F. Tuke, D. L. Dineley and B. R. Rust (1966) and are presumed to be of Palaeozoic age. The Lower Palaeozoic is thought to be conformable throughout its sequence and ranges in age from Ordovician to Lower Devonian. The lowest formation is the Allen Bay Formation which consists of several hundred metres of dolomite and dolomitic sandstone (R. G. Blackadar and R. L. Christie, 1963). This formation is overlain by the Read Bay Formation which has a thickness of 300–400 m of 'flaggy and argillaceous limestone' (Y. O. Fortier *et al.*, 1963). The youngest Palaeozoic strata are those of the Peel Sound Formation which comprises some hundreds of metres of 'boulder and pebble-conglomerate, sandstone and sandy limestone' (Blackadar and Christie, 1963). The structure of all the strata overlying the Archaean basement is simple and the area has not been subjected to any marked folding. A simplified geological map based upon the work of all the authors mentioned above is given in Figure 1.

Physiography

It is not the intention of this paper to give a regional account of the landforms. A detailed account of the terrain has been presented by J. B. Bird (1963a), who also drew attention to certain aspects of the limestone morphology (Bird, 1963b, and 1967). The area is dominated by an extensive plateau surface at an elevation of about 300 m with an amplitude of relief of about 25 m. The larger rivers are deeply incised into this plateau, often with precipitous valley sides. The lower sections of the major valleys are aggraded and typically consist of gravel plains up to 1 km or more in width. Away from the river mouths the coast is normally cliffed. The effects of glaciation are limited and there are no extensive glacial depositional features although individual erratics are common. There is abundant evidence of former high shorelines around the coast. The maximum altitude of the post-glacial marine submergence is thought to be about 50 m and marine shells found in association with the former shorelines have been radiocarbon-dated. A sample from the 'Camp Karst'[1] locality at a height of about 40 m yielded a date of 9380 ± 180 years.

Outcrops of bedrock are infrequent on the plateau surface. The surface consists of felsenmeer composed of angular limestone fragments which vary in size and shape in relation to the degree of jointing and bedding. This area is virtually devoid of soil, and D. B. O. Savile (1959, p. 96) states that '. . . it was not uncommon to see areas of up to ten acres of coarsely shattered limestone, on hill tops or upper slopes, with few lichens and not a single vascular plant'. Small patches of vegetation occur at favourable cliff-foot sites, and sedge meadow and marsh are occasionally found especially in the aggraded sections of the lower river valleys.

Climate

No detailed climatic records are available for Somerset Island, but a fully equipped meteorological station is located at Resolute Bay on southern Cornwallis Island some 80 km to the north. There is no reason to doubt that the climate of northern Somerset Island is different in any major respect from that of Cornwallis Island; for example, the depth of winter snow lying in June 1964 was very similar between the two islands. The monthly mean temperatures, precipitation as snow (converted to mm of rainfall equivalent), precipitation in the form of rain and total precipitation are given in Table I. The accumulated winter snowfall does not normally melt away completely until late June or early July and temperatures only rise above 0°C by about mid-June. Temperatures above 0°C are extremely unusual after mid-September. Northern Somerset Island therefore has a truly

TABLE I

Climatic data for Resolute Bay, 1961–66

	Jan.	*Feb.*	*Mar.*	*Apr.*	*May*	*June*	*July*	*Aug.*	*Sept.*	*Oct.*	*Nov.*	*Dec.*
Monthly mean temperature, °C	−33·2	−33·2	−30·5	−23·9	−10·8	−0·7	+3·4	+1·8	−5·6	−16·5	−22·7	−28·5
Monthly mean snow-fall, mm of water equivalent	2·8	4·0	3·3	6·6	6·9	10·5	1·8	4·4	11·5	13·0	5·1	6·4
Monthly mean rainfall, mm						3·0	24·1	20·0	3·0			
Total precipitation, mm	2·8	4·0	3·3	5·5	6·9	13·5	25·9	24·4	14·5	13·0	5·1	6·4

arctic climate with an annual average temperature of −17°C and an average annual precipitation of about 130 mm. Permafrost is continuous throughout the region and probably reaches a thickness in excess of 300 m. No evidence of any form of groundwater flow was observed.

FIELD RESULTS

Approximately 200 water samples were collected in the field during the summer of 1964 and all were analysed for their calcium and magnesium content. The analyses were all undertaken in the field, the majority within 24 hours of collection, using the titration methods initially described by G. Schwarzenbach (1957) with slight modifications as outlined in D. I. Smith and D. G. Mead (1962). The errors should be within the range of ±3 p.p.m. Temperatures were taken at the time of sample collection.

For several samples, pH values were also measured with a portable glass electrode pH meter. The pH values were obtained at the collection site and, in the case of streams issuing from snow banks, the observations were made at the point at which water emerged from under the snow. For all the pH samples a second reading of pH was obtained using the saturometer method first described by P. K. Weyl (1961) and further elaborated by R. G. Picknett (1964). This second pH value, the saturated pH reading, was obtained by adding dry, powdered 'Analar' calcium carbonate to the portion of the sample contained in the pH probe.

Samples were collected from several differing sites associated with the various stratigraphical units. It is possible to divide the samples into groups according to the nature of the collecting site.

(1) Samples of water draining from snow banks located on limestones.
(2) Samples from pools of various sizes where small patches of vegetation or sedge meadow occurred; these were underlain by limestone gravel.
(3) Samples collected from major rivers draining limestone catchments.
(4) Samples of bedrock were also analysed for their carbonate content.

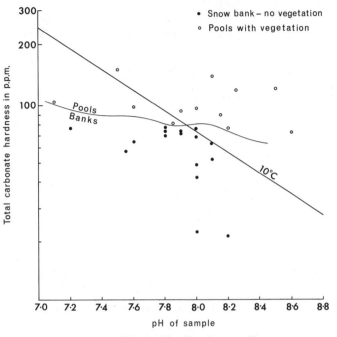

FIGURE 2. Relationship of hardness to pH

(1) Water from snow banks

Twenty-eight samples were collected from small streams and seepages issuing from snow banks. None of the snow banks was perennial. The samples were collected from four areas, namely 'Camp Karst', Cape Anne, Cunningham Inlet and the lower Hunting Valley. The hardness values for samples collected at the beginning of the melt do not appear to differ significantly from those collected later in the season. The overall mean is 46 p.p.m. $CaCO_3$ and 14 p.p.m. $MgCO_3$. In most cases there was no vegetation under the snow, but where vegetation was present, it did not form a continuous cover. Water temperatures were 0°C in every case. At sixteen of the sites, pH and saturated pH values were obtained, and the original pH value of the sample and the corresponding total hardness figure are given in Figure 2.

(2) Pools and sites associated with vegetation

At fifteen sites, samples were collected from small shallow pools that had formed on peaty soil. The peaty soil was usually only a few centimetres thick and normally covered calcareous gravel deposits; however, no limestone was exposed in the floors of the pools. The mean values were 84 p.p.m. $CaCO_3$ and 22 p.p.m. $MgCO_3$. Three of these sites had total hardness values in excess of 130 p.p.m. Water temperatures varied between 0°C and 6°C. For eleven of these sites, pH readings were obtained and these are plotted in Figure 2.

   P. W. Williams (1966) gives the results for 118 similar samples obtained from small cup depressions from the limestone outcrops in western Ireland. These are formed in bedrock but the bottoms of the pools '. . . are covered with a layer of organic and mineral matter' (p. 164). The mean values here were 66 p.p.m. $CaCO_3$ and 11 p.p.m. $MgCO_3$, with a

D.I.SMITH

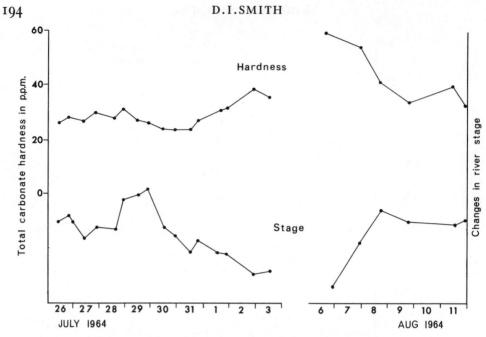

FIGURE 3. Relationship of stage to total hardness for the Hunting River

pH of 7·8. If the sampling sites are considered truly comparable, these figures are similar to those given above and suggest that rates of solution under temperate and arctic conditions are essentially similar.

(3) Major rivers

The mean values obtained for major rivers are shown in Figure 1. These sites are shown individually as it is possible to use the hardness figures to estimate the rates of erosion for individual catchments. The overall mean is 46 p.p.m. $CaCO_3$ and 11 p.p.m. $MgCO_3$. There is a relatively small range in total hardness, nine of the eleven river sites having individual means within the range 44–64 p.p.m. The Hunting River with the lowest value of 36 p.p.m. has a considerable proportion of its catchment on non-carbonate lithology, and the highest value of 95 p.p.m. is for a major river in the vicinity of 'Camp Karst'. It may be significant that this site has the highest magnesium content of the rivers sampled.

The values are in some cases based upon only single samples and it is possible that the hardness values may show fluctuations throughout the summer season. S. B. McCann *et al.* (1972) found that dissolved load concentration in rivers on Cornwallis Island increased during the flow season from 30–50 p.p.m. to 70–100 p.p.m. Twenty-four samples were collected from the Hunting River, three at the beginning of the melt season and the remaining twenty-one during the period from 26 July to 9 August. The three samples in late June gave a mean value of 34 p.p.m. $CaCO_3$ and 16 p.p.m. $MgCO_3$, while those for the later dates give a mean of 27 p.p.m. $CaCO_3$ and 7 p.p.m. $MgCO_3$. Figure 3 shows that the total hardness values appear to vary inversely with stage conditions. Facilities were not available to measure accurately the discharge of the Hunting River and the observations were made in terms of a fixed stage pole. In Figure 3 the stage conditions give therefore only a relative measure of changing discharge conditions. Figure 3 shows that the total hardness values appear to vary inversely with discharge. This is similar to the pattern found with surface limestone streams in lower latitudes (I. Douglas, 1964; D. I. Smith, F. H. Nicholson and C. J. High, 1969). Water temperatures varied from 0 to 5°C.

(4) Bedrock analyses

In order to obtain a general picture of the relationship between the calcium and magnesium composition of the bedrock and the values obtained from the water samples, fifteen rock specimens were analysed following the method described by R. E. Bisque (1961). These specimens were collected from a variety of sites in order to obtain a representative sample. The majority of the specimens were from surface debris. They were trimmed in the laboratory to exclude the weathered surface from analysis.

The results of the analysis are given in Table II. Figures are presented in terms of calcium carbonate, magnesium carbonate and acid insoluble residue. In addition, the calcium and magnesium carbonate values are given as percentages of the total carbonate value. The sites from which the specimens were collected are shown in Figure 1; at localities 3, 7, 8 and 12 more than one specimen was analysed. At locality 3 ('Camp Karst') specimens were taken from two sites on strata mapped as the Read Bay Formation and one from the Allen Bay Formation. The mean calcium and magnesium values for the water samples from all the bedrock collection sites are also given in Table II. There appears to be a general relationship between the calcium and magnesium values for the bedrock and the corresponding values from the water samples.

DISCUSSION OF RESULTS

(1) Snow banks

The first overall comment concerning the calcium and magnesium content of the water issuing from the snow banks is that the hardness values are high compared to the overall values for the area of study. The range was from a total hardness figure of 95 p.p.m. to a minimum of 18 p.p.m., but most values clustered around the means of 46 p.p.m. $CaCO_3$ and 14 p.p.m. $MgCO_3$. This value is slightly higher than the mean value for the rivers draining the limestone catchments. A further possible complication is the re-deposition of material on the underside of many of the surface pebbles, a feature common to arctic limestones and described by several workers, for example A. L. Washburn (1969) and J. C. F. Tedrow (1970). However, there is no need to invoke a hypothesis which depends on an enhanced carbon dioxide content for the atmosphere within the snow bank, for the theoretical saturation value at 0°C in terms of calcium carbonate, and using only the atmospheric carbon dioxide content, is 95 p.p.m. It must be pointed out, however, that the saturation value at 0°C is for pure calcium carbonate and some modification is necessary for rocks containing appreciable quantities of magnesium carbonate present in the form of dolomite. There is no agreement on the saturation figures that should be used for limestone of this type although, undoubtedly, dolomite and dolomitic limestone have solubilities similar in amount to that for pure calcareous limestone.

Although the hardness values obtained could be explained in terms of atmospheric carbon dioxide alone, it could also be argued that the water issuing from the snow banks is still aggressive with respect to calcium and magnesium carbonate, i.e. it has not reached its

TABLE II

Bedrock analyses and corresponding water sample analyses

Stratigraphical unit	Sample no.	Locality (Fig. 1)	Bedrock analyses % CaCO₃	% MgCO₃	% Acid insoluble residue	Bedrock analyses as percentages of carbonates % CaCO₃	% MgCO₃	Water analyses p.p.m. CaCO₃	p.p.m. MgCO₃
Peel Sound Formation	1	4	45·5	24·7	29·8	67·7	32·3 ⎫	39	10
	2	4	59·5	12·8	27·7	82·3	17·7 ⎬		
Read Bay Formation	3	3	75·0	3·4	21·6	95·7	4·3 ⎫	61	12
	4	3	77·4	5·8	17·2	93·1	6·9 ⎬		
	5	7	72·5	3·5	24·0	95·4	4·6 ⎫	48	11
	6	3	60·1	1·6	38·3	97·4	2·6 ⎬		
	7	8	78·5	2·8	18·7	96·3	3·7 ⎫		
	8	8	73·3	5·3	21·4	93·6	6·4 ⎬	48	5
	9	8	73·6	5·6	20·8	93·4	6·6 ⎭		
	10	9	78·8	6·0	15·2	92·9	7·1	54	6
	11	10	75·3	4·0	20·7	95·0	5·0	51	3
	12	12	82·7	0·8	16·5	99·0	1·0 ⎫	70	16
	13	12	81·3	0·5	18·2	99·4	0·6 ⎬		
Allen Bay Formation	14	3	36·7	30·5	32·8	54·6	45·4	64	28
	15	1	48·3	41·3	10·4	53·9	46·1	47	21

saturation value. It is possible to gain some indication of the degree of saturation of natural water by Weyl's (1961) saturometer techniques. Thus, if the artificially saturated sample has a pH greater than the original sample (is less acid) then the water is aggressive. All the samples from snow banks were either aggressive or, in a few cases, in equilibrium. None was oversaturated. The difference between the normal and artificially saturated pH values was relatively small in every case, indicating qualitatively that the samples were approaching equilibrium at the time of emergence from beneath the snow bank.

Picknett (1964) has taken this form of analysis further and has produced from theoretical and experimental data a series of diagrams which allow the degree of saturation to be expressed in percentage terms. Unfortunately these diagrams are only strictly valid for pure calcium carbonate at 10°C; the saturation line for 10°C is given in Figure 2. Nevertheless, it is worth noting that, if the observations for the snow banks in Somerset Island are plotted on these diagrams using total hardness in place of calcium hardness, all the samples are more than 90 per cent saturated and the majority have saturation figures in excess of 95 per cent. The total hardness is plotted against the natural pH of the sample in Figure 2. The equilibrium line used on this diagram is from Picknett's work for 10°C rather than the more usually quoted data which are taken from F. Trombe (1952).

The evidence therefore suggests that, on a range of limestone lithologies under the arctic climatic régime of Somerset Island, hypotheses invoking a carbon dioxide content for the snow in excess of that found in the free atmosphere are not necessary to explain the degree of solutional activity associated with snow banks. Further, the amount of solution involved in the formation of nivation hollows on limestones indicates that enrichment of the carbon dioxide content of the air in snow banks plays only a minor part in nivation processes. F. A. Cooke and V. G. Raiche (1962) have described in detail the morphology of nivation hollows in the Resolute Bay area of Cornwallis Island. These are developed on rocks of the same stratigraphical groups as the features in northern Somerset Island. They have a maximum depth of several metres; and the rates of solution indicated above could only account for the removal of a small part of the total volume of material since the area was deglaciated.

(2) Pools and sites associated with vegetation

The degree of vegetation and soil cover for samples in this group varies greatly from site to site. In nearly all cases the normal and saturated pH values show conditions of oversaturation although rarely was any form of calcium carbonate deposition observed. The mean hardness values for these sites were considerably in excess of those for water draining from snow banks, or for water from the major rivers or from pools where vegetation was not present. At three sites the total hardness values were greater than 130 p.p.m.; in all cases, the samples were taken from standing water associated with sites with a well-developed cover best described as arctic meadow. Such sites, however, are limited in areal extent. It appears that, in such soil- and vegetation-covered sites, the hardness values are significantly higher than for sites not so covered. The reasons for the increase in hardness are thought to be related to the production of carbon dioxide in the soil by biogenic processes although the direct action of humic acids may also be significant.

(3) Major rivers

The values from these sites give the best indication of the overall weathering rates for limestone in the region. Corbel (1956) outlined a method for estimating the rates of limestone solution under differing climates. The figure produced is usually expressed in terms of an overall rate of lowering. The equation suggested by Corbel was modified by P. W. Williams (1963) to take into account solutional losses in the form of magnesium and variations in rock density.

The equation is:

$$X = \frac{E(T_c + T_m)}{10D} \qquad \qquad \ldots (1)$$

where E is the effective run-off in decimetres

T_c is the mean calcium hardness expressed in p.p.m. $CaCO_3$

T_m is the mean magnesium hardness in p.p.m. $MgCO_3$

D is the density of the limestone in g/cm³

X is solution loss in mm/1000 years or m³/km²/year

The effective run-off corresponds to annual precipitation less evapotranspiration. For Somerset Island the annual precipitation in terms of rainfall equivalent is taken to be similar to that for Resolute Bay (Table I) and is 129 mm. Only estimates can be made for evapotranspiration but, using the method and the tables provided by C. W. Thornthwaite and J. R. Mather (1957), a value of 39 mm is obtained. The hardness figures represent the mean value for the eleven major rivers (Fig. 1); T_c is then 46 p.p.m. and T_m is 11 p.p.m. The density of limestone is taken to be 2·5 g/cm³.

Substitution in equation (1) gives a solution rate of approximately 2 mm/1000 years. Although many of the terms in the equation involve extrapolation and generalization, it is unlikely that the overall solution rate exceeds 3 mm/1000 years. This figure differs widely from the rate of 14 mm/1000 years given by Corbel (1959b) for a region with a periglacial climate and continuous permafrost.

CONCLUSIONS

The figures for calcium and magnesium hardness show the concentration of these solutes to be less than for comparable situations in lower latitudes, with the possible exception of the pool sites associated with organic material. These low concentration figures, considered together with the climatic data, clearly demonstrate that the rate of solution in arctic regions is small relative to that in lower latitudes. It is suggested that the major differences between arctic areas of this kind and those of lower latitudes result from the lack of a close soil and vegetation cover, so that carbon dioxide is only available from the free atmosphere. For the area studied, the enhanced values of carbon dioxide concentrations in snow banks initially suggested by J. E. Williams (1949) do not appear to be valid.

ACKNOWLEDGEMENT

The author gratefully acknowledges a grant from the University of Bristol towards the cost of illustrations. He also thanks Professor D. L. Dineley for the opportunity to join the University of Ottawa expedition to Somerset Island in 1964, and acknowledges financial assistance from the British Council.

NOTE

1. All the place names given in this paper except those marked by inverted commas have been approved by the Canadian Board on Geographical Names.

REFERENCES

ADAMS, C. S. and A. C. SWINNERTON (1937) 'The solubility of limestone', *Trans. Am. geophys. Un.* 18, 504–8

BIRD, J. B. (1963a) 'Limestone terrains in southern arctic Canada', *Proc. Permafrost int. Conf., Indiana* (Lafayette, 1963), *Nat. Acad. Sci. Publ.* 1287, 539–53

BIRD, J. B. (1963b) 'A report of the physical environment of northern Baffin Island and adjacent areas, N.W.T., Canada', *Rand Res. Mem.* 2

BIRD, J. B. (1967) *The physiography of arctic Canada*

BISQUE, R. E. (1961) 'Analysis of carbonate rocks for calcium, magnesium, iron and aluminium, with E.D.T.A.', *J. sedim. Petrol.* 31, 113–22

BLACKADAR, R. G. and R. L. CHRISTIE (1963) 'Geological reconnaissance, Boothia Peninsula, and Somerset, King William and Prince of Wales Islands, District of Franklin', *Geol. Surv. Pap. Can.* 63–19

BOLIN, B. and C. D. KEELING (1963) 'Large-scale atmospheric mixing as deduced from the seasonal and meridional variations of carbon dioxide', *J. geophys. Res.* 68, 3899–920

CLEMENT, P. and J. VAUDOUR (1967) 'Observations on the pH of melting snow in the southern French Alps' in *Arctic and alpine environments* (ed. H. E. WRIGHT and W. H. OSBURN), 205–13

COOKE, F. A. and V. G. RAICHE (1962) 'Simple transverse nivation hollows at Resolute, N.W.T.', *Geogrl Bull.* 18, 79–85

CORBEL, J. (1956) 'A new method of study for limestone regions', *Revue can. Géogr.* 10, 240–2

CORBEL, J. (1959a) 'Érosion en terrain calcaire: vitesse d'érosion morphologie', *Annls Géogr.* 366, 97–120

CORBEL, J. (1959b) 'Vitesse de l'érosion', *Z. Geomorph.* 3, 1–28

DELACOUR, F., F. WEISSEN and C. EK (1968) 'An electrolytic method for the titration of CO_2 in air', *Bull. natn. speleol. Soc.* 4, 131–6

DOUGLAS, I. (1964) 'Intensity and periodicity in denudational processes with special reference to the removal of material in solution by rivers', *Z. Geomorph.* 8, 453–73

EK, C. (1964) 'Note sur les eaux de fonte des glaciers de la Haute Maurienne, leur action sur les carbonates', *Rev. belge Géogr.* 88, 127–56

EK, C. and A. PISSART (1965) 'Dépôt de carbonate de calcium par congélation et teneur en bicarbonate des eaux résiduelles', *C. r. hebd. Séanc. Acad. Sci. Paris* 260, 929–32

FORTIER, Y. O. *et al.* (1963) 'Geology of the north-central part of the Arctic Archipelago, Northwest Territories (Operation Franklin)', *Mem. geol. Surv. Brch Can.* 326

McCANN, S. B., P. J. HOWARTH and J. G. COGLEY (1972) 'Fluvial processes in a periglacial environment: Queen Elizabeth Islands, N.W.T., Canada', *Trans. Inst. Br. Geogr.* 55, 69–82

MIOTKE, F-D. (1968) 'Karstmorphologie Studien in der glazial-überformten Höhenstufe der "Picos de Europa", Nordspanien', *Jahrb. geogr. Ges. Hannover* 4, 161 pp.

PICKNETT, R. G. (1964) 'A study of calcite solutions at 10°C', *Trans. Cave Res. Grp Gt Br.* 7, 41–62

SAVILE, D. B. O. (1959) 'The botany of Somerset Island, District of Franklin', *Can. J. Bot.* 37, 959–1002

SCHWARZENBACH, G. (1957) *Complexometric titrations*

SMITH, D. I. and D. G. MEAD (1962) 'The solution of limestone, with special reference to Mendip', *Proc. Univ. Bristol speleol. Soc.* 9, 188–211

SMITH, D. I., F. M. NICHOLSON and C. J. HIGH (1969) 'Limestone solution and the caves' in *The caves of north-west Clare, Ireland* (ed. E. K. TRATMAN)

TEDROW, J. C. F. (1970) 'Soil investigations in Inglefield Land, Greenland', *Meddr Grønland* 188 (3), 93 pp.

THORNTHWAITE, C. W. and J. R. MATHER (1957) 'Instructions and tables for computing potential evapotranspiration', *Publs Clim. Drexel Inst. Technol.* 10, 185–311

TROMBE, F. (1952) *Traité de spéléologie*

TUKE, M. F., D. L. DINELEY and B. R. RUST (1966) 'The basal sedimentary rocks in Somerset Island, N.W.T.', *Can. J. Earth Sci.* 3, 679–711

WASHBURN, A. L. (1969) 'Weathering, frost action and patterned ground in the Mesters Vig District, northeast Greenland', *Meddr Grønland* 176 (4), 303 pp.

WEYL, P. K. (1961) 'The carbonate saturometer', *J. Geol.* 69, 32–44

WILLIAMS, J. E. (1949) 'Chemical weathering at low temperatures', *Geogrl Rev.* 4, 432–44

WILLIAMS, P. W. (1963) 'An initial estimate of the speed of limestone solution in County Clare, Ireland', *Ir. Geogr.* 4, 432–41

WILLIAMS, P. W. (1966) 'Limestone pavements with special reference to western Ireland', *Trans. Inst. Br. Geogr.* 50, 155–72

RÉSUMÉ. *La dissolution calcaire dans un milieu arctique.* Peu d'observations détaillées ont été faites dans les régions polaires pour établir la forme et le degré de dissolution calcaires. La dissolution du calcaire peut y varier de celles des latitudes plus basses et les hypothèses avancées sur ce sujet sont contradictoires. La solution activée dans les régions arctique peut être dûe soit à des températures d'eau plus basses et par conséquent à un accroissement du degré de solubilité du gaz carbonique soit à une concentration particulièrement élevée du gaz carbonique de l'air sub-nivale. Une dissolution réduite pourrait être dûe à la rareté du sol et de la couverture végétale.

La région étudiée est située dans un terrain calcaire au nord-ouest de l'Île Somerset à une latitude de 74° N dans le Canada arctique. Environs 200 échantillons d'eau furent analysés sur place pour leur teneur en calcium et en magnésium. Des études limitées furent aussi entreprises pour le pH et des analyses supplémentaires furent effectuées pour les échantillons du socle rocheux. La dureté était généralement de moins de 95 mg/l et consultation des chiffres de précipitation et d'évapotranspiration donnent un degré d'érosion équivalent à 2 mm par millénaire environ. Parce que le degré de concentration des solutés et le degré d'érosion sont considérablement moins élevés que ceux des basses latitudes, on pense que l'insuffisance du sol serait le facteur significatif. On pourrait avancer avec preuves l'hypothèse que la dissolution calcaire est concentrée à la rencontre de la neige et du socle rocheux. Il n'est pas nécessaire d'invoquer une hypothèse qui ferait appel à une concentration de gaz carbonique supérieure à celle que l'on trouve normalement dans l'atmosphère libre. La région n'a pas été couverte de glace pendant ces dix derniers millénaires environ et la degré de dissolution montre qu'aucun développement important des creux sub-nivale causés par l'érosion de la dissolution n'aurait pris place à cette période.

FIG. 1. Carte géologique sommaire du nord-ouest de l'Île Somerset et degré de concentration de carbonate de calcium et de carbonate de magnésium pour les fleuves importants

FIG. 2. Graphique montrant le degré de dureté par rapport au pH

FIG. 3. Graphique montrant le niveau du fleuve par rapport du degré de dureté pour la rivière Hunting

ZUSAMMENFASSUNG. *Zur Karbonatlösung im arktischen Bereich*. Nur wenige eingehende Untersuchungen liegen bisher über Form und Ablauf der Karbonatlösung im arktischen Bereich vor. Die Karbonatlösung dort weicht möglicherweise von der in den niedrigen Breiten ab. In der Literatur finden sich die folgenden, einander widersprechenden Hypothesen: (1) eine erhöhte Karbonatlösung in der Arktis könnte durch erhöhte Kohlensäureagressivität bedingt sein, die entweder eine Folge der niedrigen Temperaturen oder des erhöhten Kohlensäuregehaltes der Luft in den tieferen Teilen der Schneedecke sei; (2) eine verringerte Agressivität könnte durch die schwache Boden- und Vegetationsentwicklung verursacht sein.

Das Untersuchungsgebiet liegt in 74° N auf Kalkstein im Nordwesten der Somerset-Insel in der kanadischen Arktis. Rund 200 Wasserproben wurden im Gelände auf Kalzium- und Magnesiumhärte sowie zum Teil auf pH-Werte analysiert. Weiterhin wurde das Gestein chemisch untersucht. Die Gesamthärte lag allgemein unter 95 mg/l $CaCO_3$. Umgerechnet auf Niederschlag und Evapotranspiration ergibt dies einen Verwitterungsbetrag von rund 2 mm/1000 Jahre. Diese Werte liegen wesentlich unter denen der mittleren und niederen Breiten. Die Abwesenheit einer geschlossenen Bodendecke mag hierfür ausschlaggebend sein. Es bestehen Hinweise zu der Annahme, dass die Karbonatlösung an den Grenzflächen Schnee/Fels besonders kräftig ansetzt. Eine Hypothese über die Anreicherung von Kohlensäure über Luftnormale wird nicht benötigt. Das Gebiet ist seit rund 10 000 Jahren eisfrei und die geringe Lösungsgeschwindigkeit berechtigt zu der Annahme, dass es in diesem Zeitraum nicht zu einer bemerkenswerten Ausbildung subnivaler Senken kommen konnte.

ABB. 1. Geologiekarte der nordwestlichen Somerset Insel und die Kalzium- und Magnesiumhärte der wichtigsten Flüsse

ABB. 2. Das Verhältnis der Gesamthärte zur pH

ABB. 3. Das Verhältnis des Stadiums zu die Gesamthärte des Hunting Flusses

Processes of solution in an Arctic limestone terrain

J. G. COGLEY

Graduate Student in Geography, McMaster University

Revised MS received 15 October 1971

ABSTRACT: Removal of material from the landscape is ultimately effected at high latitudes by the same fluvial agents as in mid-latitude regions. The snowmelt run-off season is short; it is characterized by a major spring flood and a longer period of low flow, interrupted by occasional rain-storm floods. Up to 90 per cent of annual run-off occurs during the spring flood. On Devon Island, bedload movement is confined to this period and to the rainstorm floods, and suspended sediment discharge is small except at these times. Solute concentrations vary inversely with discharge but are less variable than those of suspended sediment.

This paper describes and interprets the results of work on a small drainage basin (2.3 km²) in limestone terrain on Devon Island, N.W.T., Canada. Some emphasis is given to the rôle of running water in the solution of the limestone. The stream draining the basin flowed between 26 June and 16 August in 1970, reaching a maximum discharge of 1.42 m³/s on 2 July. Concentration of dissolved calcium and magnesium reached a maximum of 102 mg/l (as $CaCO_3$) at a discharge of 0.067 m³/s on 22 July. Calculations of stream load during a rainstorm flood indicate that solute load (Ca and Mg) was rather greater than suspended load. Solute concentration varies with the duration or distance over which solution occurs, as well as with discharge. Rainwater is a more effective solvent than old, melting snow.

GEOMORPHOLOGICAL studies of the periglacial environment have often concentrated upon specifically cryogenic landforms and processes. However, in contemporary periglacial environments such as that of the Canadian High Arctic, removal of material from the landscape is ultimately effected by the same fluvial agents as in mid-latitude regions. While cryogenic features are of considerable interest in themselves, it would be misleading to suppose that they represent the sum total of geomorphological activity in cold climatic regions.

In the Canadian High Arctic, fluvial activity assumes considerable importance alongside cryogenic activity in the suite of periglacial geomorphological processes. This paper describes and analyses fluvial activity in a small drainage basin at Radstock Bay, south-west Devon Island, Canada. The basin has been formed in limestone, and some emphasis is given in the discussion to solution processes. Although limestone solution follows predictable patterns, the results suggest that solutional activity at low temperatures is more complex in detail than the generalizations made in the past might imply.

THE BASIN OF 'JASON'S CREEK'

The stream upon which this study focuses (referred to for convenience as 'Jason's Creek') drains an area of 2·26 km² on the western side of Radstock Bay, located at 91°10′ W., 74° 40′ N., on the southern coast of Devon Island (Fig. 1). The basin consists of two distinct geomorphological units: a plateau at 280–320 m above sea level, and a sequence of raised beach deposits flanking the plateau and rising to elevations of 120 m. The stream has eroded a gully up to 150 m deep into both of these units.

The bedrock in the basin is composed of crinoidal, corallian, dolomitic and argil-

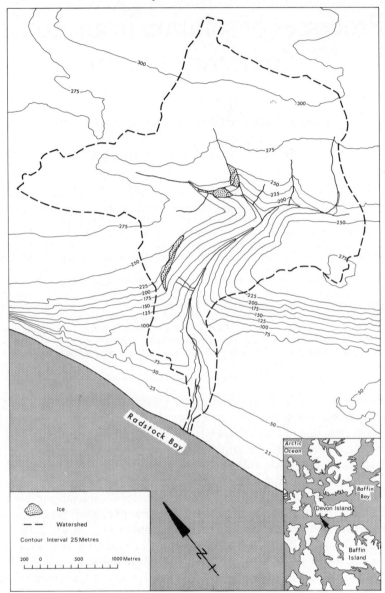

FIGURE I. Map of 'Jason's Creek', reduced from an original by P. J. Howarth. Radstock Bay is arrowed in the inset sketch at lower right

laceous limestones of the Silurian Read Bay Formation (Y. O. Fortier *et al.*, 1963). These limestones are relatively thinly bedded and dip gently north-westward. The plateau was developed on these beds in the late Tertiary. A Pleistocene ice cover which melted between 8000 and 9000 years B.P. (W. Blake, Jr., personal communication) deposited a mantle of calcareous till over the area. Events in the Holocene have been dominated by the development of the raised beach succession and by the formation of a regolith of frost-riven bedrock, solifluocted material and coarse talus on steep slopes.

Climate

The climatic conditions prevailing over south-west Devon Island are characteristically cold and dry. Mean February and July temperatures at Resolute, Cornwallis Island, 110 km to the west, are $-34 \cdot 1°C$ and $5 \cdot 1°C$ respectively; daily maxima in summer do not exceed $17°C$ (*Meterological Branch*, 1970). Mean annual precipitation at Resolute amounts to 136 mm. Forty-six per cent of this total falls as rain between late June and early September; the remainder falls as snow during winter and goes into snow-pack storage until June.

Winter in the High Arctic is characterized by the early development of an anticyclone in the west during September and October. Cold, anticyclonic air flowing over relatively warm land and water surfaces is liable to become unstable, and considerable amounts of snow fall in September and October. Precipitation is slight after the establishment of the winter anticyclone. In summer the Arctic receives frontal precipitation from depressions and, over the year as a whole, July and August are the wettest months. Table I illustrates

TABLE I

Temperature and precipitation, Resolute A Meteorological Station, 1969–70

Month	Mean daily temperature	Absolute maximum temperature	Absolute minimum temperature	Total rainfall	Total snowfall	Total snowfall (as cumulative percentage)	Total precipitation (rainfall and melted snowfall)
September	−4·7	2·2	−15·0	5·59	309·9	33·8	33·0
October	−5·4	0·0	−27·8	–	228·6	58·8	24·5
November	−26·6	−6·7	−36·6	–	38·1	63·0	3·8
December	−31·8	−14·4	−37·2	–	50·8	68·5	4·6
January	−30·5	−16·1	−41·6	–	30·5	71·9	2·8
February	−30·5	−8·3	−41·6	–	25·4	74·8	2·0
March	−29·2	−15·6	−41·1	–	30·5	78·1	2·5
April	−26·9	−13·3	−37·2	–	45·7	83·1	4·6
May	−14·2	−1·7	−27·1	–	15·2	84·7	0·8
June	−2·2	4·4	−11·1	4·32	139·7	100·0	17·5
July	4·6	16·1	−0·6	12·80	15·2	–	15·2
August	2·8	14·4	−3·9	24·2	76·2	–	32·0

Temperatures in degrees C; precipitation in mm. Note that the record for 'Jason's Creek', June–August 1970, differs somewhat from that for Resolute.

the march of temperature and precipitation at Resolute in 1969–70; the values for cumulative percentages of total snowfall indicate that two-thirds of this total had fallen by the end of November, a fact which is relevant to a discussion of solutional activity.

A further characteristic of the study area is the presence of permafrost. The mean maximum depth of the active layer for the area as a whole is between 0·5 and 0·6 m. Groundwater does not enter the hydrological cycle under these conditions, and run-off is rapid and efficient. The area is also characterized by restriction of the low-growing vegetation cover to 1–5 per cent of the terrain. This further increases the efficiency of run-off and reduces the organic contribution of CO_2 to waters dissolving the limestone.

METHODS OF DATA COLLECTION

Precipitation Precipitation was recorded at the mouth of 'Jason's Creek' in a tipping-bucket rain-gauge. Depth of precipitation as recorded on the rain-gauge charts was consistently reproducible by measurement of the rain-gauge catch in a graduated cylinder.

Stream discharge Discharge was measured with Price Type regular and pygmy current meters, and stage was recorded with a Leupold and Stevens water-level recorder. Uncertainties inherent in the calculation of cross-sectional channel area introduce errors of at least ± 5 per cent to values of discharge.

Solute concentration Dissolved calcium and magnesium were analysed in the field by complexiometric titration with E.D.T.A., using materials retailed by British Drug Houses Ltd. Most of the samples were analysed within 4 and all within 12 hours of collection. The results are accurate to ± 3 mg/l (as $CaCO_3$). Thirty samples were analysed at McMaster University for dissolved Na, K, Al and Fe, in a Perkin-Elmer 303 atomic absorption spectrophotometer; although some of these samples were 8 weeks old by the time of analysis they were stored in tightly-capped polyethylene bottles and deterioration is thought not to have been excessive.

pH Field measurements of pH were made with a battery-operated Metrohm E-280A pH meter, with manual temperature compensation. Some difficulty was experienced with sluggish meter response owing to the cold, humid conditions, and errors may range from $+0.1$ to -0.2 pH units because of this problem.

Temperature Water temperature was uniformly below $2.5°C$.

RESULTS AND DISCUSSION

Seasonal variations in discharge

The first streamflow, at a discharge of 0.107 m³/s, was observed on 26 June 1970 following a rainfall of 12 mm. Discharge remained low until 2 July, when the annual maximum discharge, 1.42 m³/s, was reached, and the stream remained in flood until about 15 July (Fig. 2). From 15 July onward discharge was relatively low except between 27 and 29 July, when the creek flooded in response to a rainfall of 26·6 mm. Figure 3 is the seasonal hydrograph of discharge for 'Jason's Creek', with graphs of solute concentration and pH superimposed.

Solid fraction of stream load

Bed load transport was confined to periods during the floods, constituting in total some 2 per cent of the flow season. Concentrations of suspended sediment were small or nil except during the snowmelt and rainstorm floods, but at these times they rose well above solute concentrations, the maximum recorded concentration of suspended sediment being 290 p.p.m. by weight. The arithmetic means of instantaneous load values for the flood of 27–29 July (obtained by multiplying discharge and concentration) were calculated as 25 g/s for suspended sediment and 31 g/s for dissolved calcium and magnesium, suggesting that solute transport during the flood was comparable with and probably somewhat greater than transport of material in suspension.

FIGURE 2. 'Jason's Creek' on 10 July. Discharge is approximately 1 m³/s

Solutional activity

The relationship of solute concentration to discharge is inverse, and non-linear in the sense that discharge increases more rapidly than concentration decreases. The basis of the relationship is the well-known dilution effect which results when the volume of water passing through a channel of relatively constant perimeter is increased: a lesser proportion of the water comes into contact with the channel perimeter and opportunities for solutional reaction become relatively fewer. The tendency for velocity to increase with discharge has an effect on the relationship, but it is fairly well described by a power curve. A series of approximations by the method of least squares on the data collected in 1970 produced the following equations for the curve:

$$Ca = 42 \, Q^{-0.118}, r = -0.797 \tag{1}$$

$$(Ca + Mg) = 56 \, Q^{-0.118}, r = -0.825 \tag{2}$$

Q represents discharge in m³/s, and Ca and (Ca + Mg) represent concentrations of solute (as $CaCO_3$) in mg/l; the number of observations was 72, and values of r were statistically significant at the 0·01 per cent confidence level. Ca and (Ca + Mg) are plotted in Figure 4A and B respectively.

The equations were derived for waters sampled at the mouths of 'Jason's Creek' and of the Mecham River, south-east Cornwallis Island. The Mecham River is an appreciably larger stream than 'Jason's Creek', but their respective basins are eroded into the same lithological units and are physiographically rather similar.

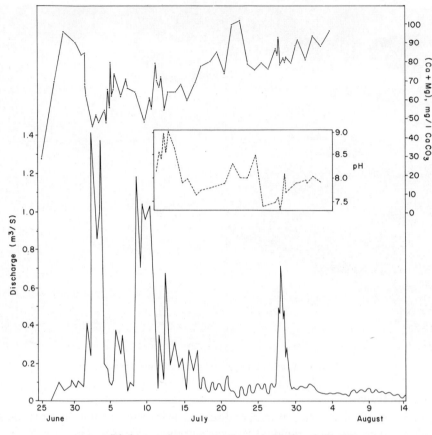

FIGURE 3. Discharge, solute concentration and pH, 'Jason's Creek', 1970

Five other sample groups are plotted on Figure 4. These include samples from standing water, from rills draining soliflual and talus slopes, and from 'Jason's Creek' during the
flood of 27–29 July, together with several sets of rill samples taken in sequence downstream.
It must be stressed that the location of the rill and standing water samples is entirely
arbitrary with reference to the discharge axes of the graphs, serving illustrative purposes
only.

The highest concentrations among all the samples in Figure 4 are found in samples
from standing waters. This observation and the general form of the solute–discharge graph
suggest that a relationship exists whereby solute concentration becomes a function of the
duration or distance over which solution occurs, probably approaching some limiting value
as an asymptote. The results of the downstream sample sequences indicate that rates of
solution are initially rapid, for the downstream sequences extend over no more than 150 m.
Most extend over less than 100 m. If these sets of samples are taken to belong to the same
population as the other rill samples, and if the suggested asymptotic relationship between
solute concentration and length of contact holds true, it appears that solution by rill waters
is substantially completed in the first 100–150 m of flow.

For this relationship to remain valid, initial conditions and bounding conditions must

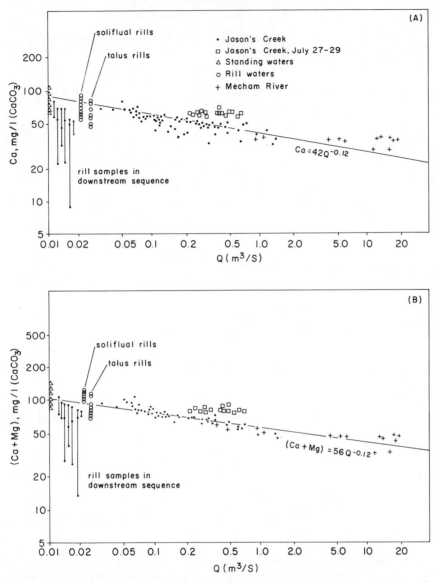

FIGURE 4. Plot of discharge against (A) dissolved Ca, (B) dissolved (Ca+Mg), for samples collected on Devon and Cornwallis Islands, 1969–70. Solute concentrations as CaCO₃

remain constant. In this context, discharge, for instance, may be thought of as a bounding condition. If discharge increases, the dilution effect expressed by equations (1) and (2) comes into operation, reducing the rate of solution per unit volume of solvent.

The importance of initial conditions is particularly apparent in the search for an explanation of the anomalous features noted during the flood of 27–29 July. The concentrations recorded during this flood manifested no fluctuation with discharge, save for an initial increase to 91 mg/l CaCO₃ of (Ca + Mg). For the remainder of the flood, solute concentra-

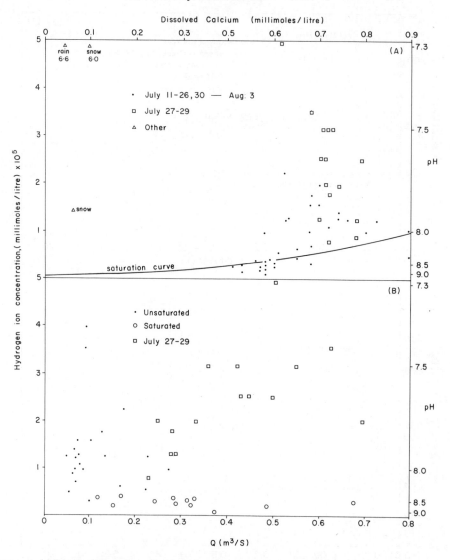

FIGURE 5. (A) Dissolved Ca against pH, (B) discharge against pH, 'Jason's Creek', 1970. Note (i) 1 millimole per litre ≃ 100 mg/l CaCO₃; (ii) supersaturated samples lie below the curve in Figure 5A. The pH axes of the graphs have been changed to amplify visually the contrast between groups of samples

tion was constant at or near 80 mg/l CaCO₃. This failure of the solute-discharge relationship may be explained by considering the initial ionic constitution of the waters reaching solutional surfaces. Figure 5A is a plot of dissolved Ca against pH for samples taken at the mouth of 'Jason's Creek', together with two samples of meltwater dripping directly from snowpatches, collected on 12 July and 1 August, and one sample from the rainwater which fell on 28 July. The first snowmelt sample had a pH of 7·85 and Ca concentration of 6 mg/l CaCO₃; the corresponding values for the rainwater were 6·0 and 6. The pH of the rainwater was almost two pH units below that of the meltwater. The second meltwater sample had a pH of 6·6 and Ca concentration of 9 mg/l $CaCO_3$; it is thought, although it cannot be conclusively demonstrated, that the lower pH of this sample resulted from dilution of the meltwater proper by rainwater.

The solutional behaviour of the rainwater appears to be quite different from that of the meltwater. Within the range of Figure 5A the Ca concentration of the rainwater run-off is unrelated to pH, whereas for the meltwater samples a correlation coefficient of $-0·592$ (significantly different from zero at the 0·01 level with fifty observations) was found between the two. A saturation curve has been drawn on Figure 5A using R. G. Picknett's values (1964) for the solubility products at 10°C; all of the samples were colder than 10°C, but at 0°C the upward translation of the curve would be such that a maximum of two of the unsaturated samples would be displaced to the supersaturated side of the curve. None of the rainwater samples falls below the curve, while supersaturation with respect to calcite is indicated for several of the meltwater samples.

All but two of these apparently supersaturated samples were collected between 11 and 13 July. No sound explanation can be offered for this fact. Figure 5B, however, indicates that these samples occur in general at higher discharges than the unsaturated snowmelt samples. This explains why the supersaturated samples contain less Ca than unsaturated waters, but not why the waters which contain less Ca should be supersaturated rather than unsaturated.

The disparity between the pH values for meltwater and for rainwater, and the apparent solutional results of this disparity, probably reflect a disparity in the concentration of 'free' CO_2. The availability of aqueous CO_2 and its derivatives is more significant in limestone solution than is the activity of the hydrogen ion, and it is reasonable to postulate a good correlation between the two.

C. Ek (1964, 1966) found the pH of melting glacier ice in the Savoy Alps and the Italian Dolomites to lie between 7·65 and 8·9, decreasing downstream and up-glacier from the glacier snout; values as low as 7·35 were found in the firn zone. He attributed high pH in melting ice to the expulsion of gaseous CO_2 from the ice upon densification. P. Clement and J. C. Vaudour (1968), in contrast, report pH values ranging from 4·4 to 7·0 about a mean of 5·4 for melting snow in the southern French Alps. Their observations were made in winter, while Ek's were made in autumn (September). From the work of Ek and of Clement and Vaudour, and from the results of work on 'Jason's Creek', it is possible to predict that the pH of snow increases with depth in the snowpack. The implication of this prediction for solution studies is that increasing pH corresponds to decreasing CO_2 concentration, and thus that the aggressiveness of snow meltwater towards calcite decreases as the snow ages. In this respect one would expect confirmation of D. C. Ford's findings (1971) of low CO_2 concentrations in glacial meltwaters above the treeline in the Canadian Rocky Mountains; these waters saturate with respect to calcite at Ca concentrations of less than 50 mg/l (as $CaCO_3$).

Lesser ions in solution

Concentrations of Na, K, Al and Fe as determined by spectrophotometry were all relatively small. The samples were collected between 11 July and 16 August. No temporal trends were detectable in the concentrations of any of these ions; Na ranged from 1·66 to 3·96 mg/l, K from 0·25 to 0·8 mg/l, while Al and Fe were either not detectable or present only in trace quantities.

ACKNOWLEDGEMENTS

S. B. McCann made available funds, awarded to him by the National Research Council of Canada, for this research. P. J. Howarth made the original map from which Figure 1 was taken. Some of the results for Ca and Mg concentration used in this paper were collected by S. B. McCann from the basin of 'Jason's Creek' and immediately adjacent areas in 1969, and by S. B. McCann and P. J. Howarth in 1970 from the Mecham River, Cornwallis Island. Their permission for the inclusion of these results is gratefully acknowledged.

REFERENCES

CLEMENT, P. and J. C. VAUDOUR (1968) 'Observations on the pH of melting snow in the southern French Alps' in *Arctic and alpine environments* (ed. H. E. WRIGHT, Jr. and W. H. OSBURN), 205–13

EK, C. (1964), 'Note sur les eaux de fonte des glaciers de la Haute Maurienne: leur action sur les carbonates', *Rev. belge Géogr.* 88, 127–56

EK, C. (1966) 'Faible aggressivité des eaux de fonte des glaciers; l'exemple de la Marmolada (Dolomites)', *Annls Soc. géol. Belg.* 89, 117–88

FORD, D. C. (1971) 'Characteristics of limestone solution in the southern Rocky Mountains and Selkirk Mountains, Alberta and British Columbia', *Can. J. Earth Sci.* 8, 585–609

FORTIER, Y. O., *et al.* (1963), 'Geology of the north central part of the Arctic Archipelago', *Mem. geol. Surv. Brch Can.* 320

Meteorological Branch, Department of Transport, Government of Canada (1970), 'Climate of the Canadian Arctic', *Can. Hydrogr. Serv., Marine Sci. Br., Dept. Energy, Mines and Resources* (Ottawa)

PICKNETT, R. G. (1964) 'Calcite solutions at 10°C.,' *Trans. Cave Res. Gp Gt Br.* 7, 36–62

RÉSUMÉ. *Les processus de solution dans un terrain calcaire dans l'Arctique.* Le transport des matériaux de dénudation jusqu'à la mer est effectué dans les régions polaires par le même mécanisme, les fleuves, comme dans les régions du sud. La saison du débit de fonte de neige est courte et consiste en une pointe au printemps, suivie par une période plus longue de débit moins fort. Ce régime est interrompu de temps en temps par des montées d'écoulement dues aux pluies. La pointe du printemps contribue jusqu'à 90 pour cent du débit annuel total. Sur l'Île Devon, le transport de sédiments roulant dans le lit n'a lieu que pendant les périodes de montée et le transport de sédiments en suspension est significatif seulement durant ces périodes. Les concentrations de matières en solution sont en rapport inverse avec les débits, mais elles sont moins variables que celles des sédiments en suspension.

Cet article présente, avec interprétation, les résultats d'une étude d'un petit bassin d'écoulement (superficie de 2,3 km²) dans un terrain calcaire sur l'Île Devon, T.N.O., Canada. L'auteur met l'accent sur le rôle des eaux mouvantes dans la solution du calcaire. En 1970, le ruisseau qui draine le bassin fut en mouvement du 26 juin jusqu'au 16 août et un débit maximum de 1,42 m³/s fut atteint le 2 juillet. La plus grande concentration de calcium et de magnésium en solution, soit 102 mg/l (sous forme de CaCO₃), fut observée, à un débit de 0,067 m³/s, le 22 juillet. Des calculs du transport de matières par le ruisseau lors d'une montée d'écoulement de pluie indiquent que la contribution au transport total des matières en solution (Ca et Mg) était plus grande que celle des matières en suspension. Les concentrations de matières en solution varient avec la durée du processus de solution, ou avec la distance sur laquelle il a lieu, ainsi qu'avec les débits. La pluie est plus efficace dans le processus de solution que l'eau de fonte de la vieille neige.

FIG. 1. Carte du « Jason's Creek », réduite d'une carte originale par P. J. Howarth. Radstock Bay est indiqué par une flèche dans le médaillon de droite, en bas

FIG. 2. Vue de « Jason's Creek », le 10 juillet. Le débit d'eau est app. 1m³/s

FIG. 3. Débit d'eau, concentration de matières en solution et pH, « Jason's Creek », 1970

FIG. 4. Graphiques du débit d'eau et (A) Ca en solution, (B) Ca+Mg en solution, pour les eaux étudiées sur l'Île Devon et l'Île Cornwallis, 1969–70

FIG. 5. (A) Ca en solution et pH, (B) débit d'eau et pH, « Jason's Creek », 1970. Notez, (i) 1 millimole / litre ≃ 100 mg / l CaCO₃; (ii) l'échantillonage des eaux supersaturées est indiqué au bas de la courbe de Fig. 5A. Les axes illustrant le pH ont été modifiés pour amplifier le contraste visuel entre les groupes d'échantillons

ZUSAMMENFASSUNG. *Lösungsprozesse in einem arktischen Kalkgelände.* Die Verschiebung des Materials von der Landschaft hinaus führt in den Polarzonen durch dieselben fluvialen Mechanismen die in den gemässigten Zonen wirken. Die kurze Jahreszeit, während welcher das Schneeschmelzwasser abläuft, besteht aus einem Abflusshöchststand im Frühling mit einer längeren nachfolgenden Periode geringeren Abflusses, die von gelegentlichen Regenwasserüberschwemmungen unterbrochen wird. Bis zu 90 Prozent des jährlichen Abflusses läuft im Frühling ab. Auf der Devon Insel fällt die Beförderung des Flussbettmaterials nur in diese Jahreszeit und in den Regenwasserüberschwemmungsperioden; schwebendes Sediment wird in grossen Mengen nur zu diesen Zeiten getragen. Gelöste Sedimentkonzentrationen wechseln umgekehrterweise mit dem Abfluss, sind aber nicht so veränderlich wie schwebende Sedimentkonzentrationen.

Dieser Aufsatz beschreibt und erklärt das Verhalten eines kleinen Wasserbeckens (Bodenfläche 2,3 km²) auf einem Kalkgelände auf der Devon Insel, N.W.G., Kanada. Man legt Nachdruck auf die Rolle fliessenden Wassers in der Kalklösung. Der Bach, der das Becken entwässert, floss zwischen 26. Juni und 16. August 1970 und erreichte bis einen maximalen Abfluss von 1,42 m³/s am 2. Juli. Die Konzentration gelösten Kalziums und Magnesiums erreichte ein Maximum von 102 mg/l (als CaCO₃) bei einem Abfluss von 0,067 m³/s am 22. Juli. Eine Berechnung der Last des Baches für eine Regenwasserüberschwemmungsperiode deutet an, dass die gelöste Sedimentmenge (Ca+Mg) etwas grösser als die schwebende Sedimentmenge war. Gelöste Sedimentkonzentrationen wechseln mit der Dauer und der Länge des Lösungsprozesses sowohl as mit der Abflussmenge. Das Regenwasser ist ein wirksameres Lösungsmittel als alter schmelzender Schnee.

ABB. 1. Karte von ‚Jason's Creek', verkleinert von eine Urkarte von P. J. Howarth. Ausschnitt, rechts unten: ein Pfeil zeigt Radstock Bay an

ABB. 2. ‚Jason's Creek' am 10. Juli. Abfluss *ca.* 1 m³/s

ABB. 3. Abfluss, lösliche Konzentration und pH, ‚Jason's Creek', 1970

ABB. 4. Graphische Darstellung von Abfluss gegenüber (A) gelösten Ca, (B) gelösten (Ca+Mg), für Messungen auf der Devon Insel und der Cornwallis Insel, 1969–70

ABB. 5. (A) Gelöstes Ca gegenüber pH, (B) Abfluss gegenüber pH, ‚Jason's Creek', 1970. Bemerkung: (i) 1 Millimol/Liter ≃ 100 mg/l CaCO₃; (ii) die übersättigte Werte liegen unter der Kurve von Abb. 5A. Die pH-Achsen der Diagramme wurden modifiziert, um den Gegensatz zwischen Messungsgruppen besser sichtbar zu machen.

The British Geomorphological Research Group

The British Geomorphological Research Group was founded in January 1961 to encourage research in geomorphology, to undertake large-scale projects of research or compilation in which the co-operation of many geomorphologists is involved, and to hold field-meetings and symposia. In January 1970 the Group was constituted a formal Study Group of the Institute of British Geographers. Founded by the late Professor David L. Linton, it has a membership of over 200; its publications include a series of Technical Bulletins, a Register of Current Research in Geomorphology, and a Bibliography of British Geomorphology.

Details of membership and lists of publications for sale may be obtained from the Hon. Secretary of the Group, Dr E. Derbyshire, Department of Geography, The University of Keele, Keele, Staffs., ST5 5BG.

The Institute of British Geographers

Details of Membership are available from the Administrative Assistant, Institute of British Geographers, 1 Kensington Gore, London, S.W.7.

Papers or monographs intended for publication must be sent in the first place to the Hon. Editor, Dr C. Embleton, Department of Geography, King's College, Strand, London, W.C.2. Papers for reading at the Annual Conference (even if they are subsequently to be considered for publication) should, however, be sent to the Hon. Secretary, Professor R. Lawton, Department of Geography, University of Liverpool.

Requests for copies of publications and separate papers should be made to the Administrative Assistant, who also has available 'Notes for the guidance of authors submitting papers for publication by the Institute'.

Publications in Print

TRANSACTIONS AND PAPERS

No. 11. 1946. £1.50 *net*
No. 13. 1947. £1.25 *net*
No. 14. 1948. £1.25 *net*
No. 15. 1949. £2.25 *net*
No. 16. 1950. £1.50 *net*
No. 17. 1951. £2.25 *net*
No. 18. 1952. £1.75 *net*
No. 19. 1953. £1.75 *net*
No. 20. 1954. £2.00 *net*
No. 21. 1955. £1.50 *net*
No. 22. 1956. £1.50 *net*
No. 23. 1957. £2.00 *net*
No. 25. 1958. £2.00 *net*
No. 26. 1959. £2.00 *net*
No. 28. 1960. £2.00 *net*
No. 29. 1961. £2.00 *net*
No. 30. June 1962. £1.25 *net*
No. 31. December 1962. £1.25 *net*
No. 32. June 1963. £1.25 *net*
No. 33. December 1963. £1.75 *net*
No. 34. June 1964. £1.75 *net*
No. 35. December 1964. £1.75 *net*

TRANSACTIONS

No. 36. June 1965. £1.75 *net*
No. 37. December 1965. £1.75 *net*
No. 38. June 1966. £2.25 *net*
No. 40. December 1966. £2.50 *net*
No. 41. June 1967. £2.50 *net*
No. 42. December 1967. £2.50 *net*
No. 44. May 1968. £2.50 *net*
No. 45. September 1968. £2.50 *net*
No. 46. March 1969. £2.50 *net*
No. 47. September 1969. £3.00 *net*
No. 48. December 1969. £3.00 *net*

No. 49. March 1970. £3.00 *net*
No. 50. July 1970. £3.00 *net*
No. 51. November 1970. £3.00 *net*
No. 52. March 1971. £3.00 *net*
No. 53. July 1971. £3.00 *net*
No. 54. November 1971. £3.00 *net*
No. 55. March 1972. £4.00 *net*
No. 56. July 1972. £4.00 *net*

TRANSACTIONS: MONOGRAPHS

No. 3. *The changing sea-level* by H. Baulig (1935). £0.60 *net*
No. 12. *Some problems of society and environment* by H. J. Fleure (1947). £0.60 *net*
No. 24. *The human geography of southern Chile* by G. J. Butland (1957). £1.00 *net*
No. 27. *Alnwick, Northumberland: a study in town-plan analysis* by M. R. G. Conzen (1960). £2.25 *net*
No. 39. *The vertical displacement of shorelines in Highland Britain.* (Transactions, October 1966). £2.00 *net*
No. 43. *Population maps of the British Isles, 1961.* (Transactions, April 1968). £1.20 *net*

SPECIAL PUBLICATIONS

No. 1. *Land use and resources: studies in applied geography. A memorial volume to Sir Dudley Stamp* (November 1968). £3.50 *net*
No. 2. *A geomorphological study of postglacial uplift with particular reference to Arctic Canada* by J. T. Andrews (1970). £2.50 *net*
No. 3. *Slopes: form and process* (January 1971). £3.00 *net*

These publications may be obtained from any bookseller at the prices quoted above. Members of the Institute may buy copies at two-thirds of these prices. Application for these should be made to the Administrative Assistant, Institute of British Geographers, 1 Kensington Gore, London, S.W.7, who also has available for sale a number of copies of *Problems of Applied Geography*: Proceedings of the Anglo-Polish Seminar, 1959 (Warsaw, 1961) at £1.25; *Problems of Applied Geography II*: Proceedings of the Anglo-Polish Seminar, Keele, 1962 (Warsaw, 1964) at £3.00.